DATE DUE

OCT - 4 1994 5316	
MAR. 2 8 1996	

The Economics of Solid Waste Reduction

New Horizons in Environmental Economics

General Editor: Wallace E. Oates, Professor of Economics,
University of Maryland

This important new series is designed to make a significant contribution
to the development of the principles and practices of environmental
economics. It will include both theoretical and empirical work.
International in scope, it will address issues of current and future concern
in both East and West and in developed and developing countries.

The main purpose of the series is to create a forum for the publication of
high quality work and to show how economic analysis can make a
contribution to understanding and resolving the environmental problems
confronting the world in the late 20th century.

Innovation in Environmental Policy
Edited by T. H. Tietenberg

Environmental Economics
Policies for Environmental Management and Sustainable Development
Clem Tisdell

The Economics of Solid Waste Reduction
Robin R. Jenkins

The Economics of
Solid Waste Reduction

The Impact of User Fees

Robin R. Jenkins
Assistant Professor of Economics
Saint Mary's College of Maryland

Edward Elgar

Published by
Edward Elgar Publishing Limited
Gower House
Croft Road
Aldershot
Hants GU11 3HR
England

Edward Elgar Publishing Company
Old Post Road
Brookfield
Vermont 05036
USA

A CIP catalogue record for this book is available from the British Library

A CIP catalogue record for this book is available from the US Library of Congress

ISBN 1 85278 673 6

Printed and Bound in Great Britain by
Hartnolls Limited, Bodmin, Cornwall.

Contents

viii   *Contents*

Tables

Figures

Preface

Readers interested in the practical issues concerning user fees for solid waste services (SWS) should find certain chapters of this book particularly helpful. Such readers should start with Chapter 1 which gives a straightforward synopsis of the entire book. It describes the failure currently occurring in the market for SWS and presents an overview of the research undertaken to analyze the effectiveness of a user fee in correcting that market failure. Other chapters that are user friendly and address practical issues are Chapters 2, 5, 6, 8 and 9. For a description of the contents of each of these chapters, see the final section of Chapter 1.

The model developed in the book forecasts the quantity of waste discarded by a particular community and the impact of a user fee on that quantity of waste. Chapter 8 gives easy-to-follow instructions for readers interested in forecasting.

Three chapters are highly technical and will only be useful to economists and perhaps other social scientists or engineers. These are Chapters 3, 4 and 7, which describe the theoretical model, the empirical model and tests of the empirical model, respectively.

Most of the research undertaken is related to US data. However, the results may apply to any community with a scarcity of disposal capacity. This describes many European as well as American communities. The results, at the very least, should lend insight into how effective a user fee for residential SWS can be in correcting the failures in such markets.

Acknowledgements

This book grew out of my PhD dissertation and I extend a heartfelt thank you to Dr Harry H. Kelejian, my dissertation advisor. He devoted many hours to the empirical issues that arose in this project, and it was his expertise that permitted the use of sophisticated econometric techniques. Dr Kelejian also carefully reviewed the dissertation for clarity of presentation. Detailed editing is rare among accomplished professors with strained schedules, yet it is exactly those professors whose opinions are most valuable. I also thank Dr Kelejian for his consistent availability and for the many enlightening discussions that shaped and directed the dissertation. Finally, I thank him for the direct encouragement he gave me and, just as importantly, for the contagious excitement with which he approached this project.

I thank Dr Robert M. Schwab for his careful review of the theoretical chapter and for his suggestion that I create the annual model. I am grateful, also, for Dr Schwab's quick insight and his fresh, commonsense perspective, which have on more than one occasion turned complicated tangles into manageable tasks.

Many thanks to Dr Maureen L. Cropper for suggesting that I investigate the impact of user fees on residential demand for solid waste services as a dissertation topic. Her comments and suggestions throughout this project were invaluable.

Dr Wallace E. Oates gave many helpful suggestions regarding the organization of the book. I am indebted to him also for suggesting the stand-alone chapter that explains how to use the model for forecasting.

The forecasting technique presented in Chapter 8 was improved by a suggestion from Charles Trozzo of Charles Rivers Associates in Washington, DC.

I am indebted to Howard Levenson of the Office of Technology Assessment for getting me started on this project by sharing information and discussing the solid waste problem. Many thanks to Kathy Frevert of the Environmental Protection Agency, Clark Wiseman of Resources for the Future, and Lisa Skumatz and Jennifer Bagby of the Seattle Solid Waste Utility for their suggestions and discussion time.

I could not have compiled my data set without the municipal personnel who patiently explained the data they shared with me. In particular, I thank Jody Pospisil of Estherville City Hall, Carol Krueger of the City of Albuquerque, James J. Palmer of the Hillsborough County Board of Commissioners, Claire R. Knapp of the Borough of High Bridge, Elmer Hite of the City of Spokane, Steven Hudgins of the Howard County Department of Public Works, Dan Morgan of the City of St. Petersburg, and Joe Johnson and Monique Albert of San Francisco.

Finally, I thank my brother, Duane Jenkins, for expert computer assistance. And, I thank my parents, Joe and Gloria Jenkins, and especially my husband, Eric Kuhl, for their unfailing interest and encouragement.

1. Introduction

1.1 THE PROBLEM WITH THE MARKET FOR SOLID WASTE SERVICES

Most households in the United States are either charged a flat fee to have their refuse collected or they pay for it through property taxes. In both cases the household's payment is fixed regardless of the quantity of refuse discarded. Households face a marginal or incremental cost of refuse disposal equal to zero, and thus they may be disposing of greater than optimal quantities of waste.

The fee the household pays for refuse collection must cover the cost of transporting and disposing of the waste as well as of collecting it. Thus, we shall henceforth talk about the household's demand for solid waste services (SWS) rather than for refuse collection. SWS include the collection, transportation and disposal of solid waste.

Consider the demand curve for residential SWS shown in Figure 1.1. As with virtually any demand curve, as price declines, demand for the good increases. In the case of SWS, the price per unit in many communities is zero. Thus the quantity of SWS demanded, measured as the quantity of waste discarded, is Q^z. This is more than the quantity that would be demanded if the going price for SWS were positive. In Figure 1.1, if a positive price, P*, were charged per unit of SWS - for example, per 32-gallon container of refuse - the quantity of SWS demanded would decline to Q*.

For the quantity of waste not only to decline but also to be optimal, the price charged for SWS should reflect the social cost of such services. By social cost we mean the sum of the private and public costs of SWS. An example of a private cost is the cost of the land on which a landfill is built. An example of a public cost is the expected cost of potential groundwater contamination from a landfill. To reduce waste to its optimal level, the price per 32-gallon container of SWS should equal the social cost of discarding those 32 gallons. In other words, the price should reflect the marginal social cost of SWS. In Figure 1.1 we assume that the marginal social cost is P*. Thus Q* is the optimal quantity of SWS.

1

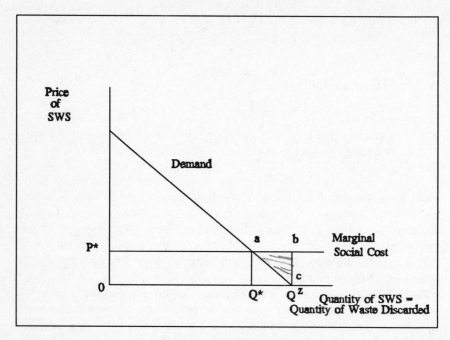

Figure 1.1 The Demand Curve for Residential SWS

Consider the quantity of waste discarded in Figure 1.1 beyond the optimal quantity - from Q^* to Q^z. The cost to society of disposing of such waste is greater than the benefit to households of having it discarded.[1] This disparity leads to a welfare loss to society equal to the area of the triangle, abc.

The market for SWS is not a special case. Any good whose price is zero is likely to be over-consumed to a point where society suffers a welfare loss. Here we have a service, SWS, whose price is zero which leads households to over consume it.

Two possibilities might quiet our concern over the prevalence of zero incremental pricing for SWS. The first is that the marginal social cost of SWS is close to zero. If so, then the welfare loss caused by a zero incremental price might be negligible. In Figure 1.1, if the marginal social cost curve is close to the horizontal axis, clearly the welfare loss triangle becomes small. The second possibility is that the demand curve for SWS is nearly vertical. This depends entirely on the responsiveness of households to a user charge for SWS. By a user charge we mean a fee based on the volume or weight of waste discarded; in other words, a price per unit of SWS. If households are not responsive to a user charge, the demand curve will be nearly vertical and once again the welfare loss caused by a zero price for SWS should be negligible, since Q^z will be very close to Q^*. We shall consider each of these issues in turn.

1.11 The Marginal Social Cost of SWS

Recent concern over the lack of incremental pricing for SWS has probably been generated by what many US communities consider a solid waste crisis. The Environmental Protection Agency (EPA) has estimated that 80 percent of existing landfills will reach capacity and shut down in the next 20 years (US EPA, 1988, p. 10). These closures coupled with a decline in the rate at which new landfills are being opened are responsible for the crisis in landfill space, or more precisely a scarcity of such space. For example, there were 381 new landfills opened in 1970 but only 62 in 1986 (US EPA, 1988, P II-13).

Three factors have contributed to the shortage of landfill space. The most obvious is the decreased availability of inexpensive land in the crowded northeastern US or in any densely populated region. On average, a landfill facility requires 86.5 acres.[2]

Another important factor is the public's increased awareness of the negative by-products of disposal sites. Besides the odors and truck noises which accompany disposal sites, many sites pose serious environmental threats. Landfills, for example, have been known to contaminate ground and surface waters. In addition they generate methane gas, which can migrate underground to the basements of nearby homes. When trapped in a basement, methane gas can concentrate and become explosive.

Although landfills are certainly our primary concern, since approximately 75 percent of municipal waste in the US is discarded in them,[3] other disposal facilities pose their own environmental threats. Incinerators, or waste-to-energy facilities - both of which are quite popular in parts of Europe - emit fly ash, which contains pollutants. They also generate bottom ash, whose level of toxicity is the subject of an ongoing controversy.

In short, there is good evidence that the public's concerns about disposal facilities are sound. As a final bit of evidence, consider that over 20 percent of the sites on the Super Fund Priority List for Hazardous Waste Sites are old solid waste dumps (US Congress, Office of Technology Assessment, 1989, p. 42). Of course, a result of this public awareness and aversion is that new disposal sites have become quite expensive and time consuming to locate. Recent estimates of the number of years required to site a landfill - that is, simply to find and secure the land on which to build - range from two to seven years.[4] An extreme example is Sweden, where solid waste is typically incinerated. A moratorium on new incineration plants was imposed there from 1984 to 1986 (Platt, et al., 1988, p. 4).

To further exacerbate the scarcity of disposal capacity in the US, soon landfills and incinerators will be governed by stricter EPA regulations. Although these regulations have yet to be passed, many states use the proposed regulations as guidelines for current construction projects. The proposed regulations include requiring a landfill liner of either natural or synthetic materials. The liner is to guard against contaminated water (leachate) seeping from the landfill into ground water. Other proposed requirements are installation of methane gas collection systems and consistent monitoring of nearby ground and surface waters. Thus disposal site construction and maintenance has become more expensive than ever before.

Europe also is experiencing a shortage of disposal capacity. In the countries that rely heavily on landfilling, the reasons for the shortage are very similar to the reasons in the US. The countries like Sweden that rely heavily on incinerating waste faced high population densities and high land prices decades ago and, as a result, switched from landfilling as the primary method of disposal to incinerating. Table 1.1 gives the percent of the waste stream, after recycling, that is landfilled in 11 countries. Even those countries who rely primarily on incinerating are experiencing a disposal capacity shortfall, because the environmental hazards associated with incinerating were not widely recognized until the early 1980s. After that the incinerator siting process became difficult and expensive.

The high value of land, the public's aversion to living near disposal sites and expensive new environmental regulations have all increased the private cost of SWS. Although public costs of SWS, such as the expected cost of ground water contamination or methane gas migration, may have been high in the past, we have only recently become aware of them. Recall that the social cost of SWS is the sum of its private and public costs. Thus, the marginal social cost of SWS seems quite significant. Setting the price per unit of SWS equal to zero is exacerbating the scarcity of landfill space. Households are not being given the proper incentive to reduce the quantity of waste they discard to its optimal level - Q* in Figure 1.1.

1.12 The Responsiveness of Households to User Fees for SWS

Several communities have acknowledged the possible inefficiencies of flat fee pricing and are now charging user fees or an incremental price for SWS. However many policy makers argue against user fees, claiming that a user *fee would be too small an item in the overall household budget to cause*

Table 1.1 The Percent of Solid Waste, After Recycling, That Is Landfilled in the US and Select European Countries

Country	Percent Landfilled*	Year
Denmark	44	1985
France	54	1983
Greece	100	1983
Ireland	100	1985
Italy	85	1983
Netherlands	56-61	1985
Sweden	35-49	1985, 1987
Switzerland	22-25	1985
United Kingdom	90	1983
United States	90	1986
West Germany	66-74	1985, 1986

*Percentage of the total weight of commercial and residential solid waste that is landfilled after deducting the amount recycled.

Source: US Congress, Office of Technology Assessment. (1989) *Facing America's Trash: What Next for Municipal Solid Waste*, US Government Printing Office, Washington, DC, October, p. 273.

people to change their general consumption patterns. Thus, solid waste quantities will not respond to a user fee, since consumption activities are the source of all residential solid waste. These policy makers are suggesting that the demand curve for SWS is nearly vertical or inelastic. If one replaces the demand curve which appears in Figure 1.1 with a vertical one, clearly the welfare loss associated with a zero price for SWS disappears. Notice, however, that policy makers focus on the quantity of waste *generated* by households. Certainly a user fee might have only small effects on the quantity of waste households *generate*. However, a user fee can easily reduce the quantity of waste actually *discarded* by affecting the amount of waste households recycle.

In sum there are three ways households can reduce the quantity of waste they discard in response to a user fee. The first is to reduce the quantity of waste generated by avoiding heavily packaged items and short-lived products. We suspect this response is rather small. The second, probably larger, response is to recycle certain waste materials. The third response to a user fee is a negative one. Households will have incentive to discard waste by littering and illegally dumping their refuse. Often, in the user fee communities we have studied, illegal dumping takes the form of placing refuse in commercial waste receptacles. To prevent this, the user fee communities require households to subscribe to a minimum service level. For example, they require each household to subscribe to one can of service per week. The implicit assumption is that no household can completely eliminate its waste.

The three ways that households can reduce waste in response to a user fee for SWS are three reasons to suspect that the demand curve for SWS is not vertical. When the price of SWS increases, households can respond primarily by recycling. Perhaps the most important goal of our research is to provide an empirical estimate of the demand curve for SWS. Our results suggest that households will respond to a positive fee per unit of SWS.

1.13 Household versus Commercial Waste

So far we have only discussed household waste, but commercial waste also goes to municipal solid waste landfills. In fact, the EPA defines a municipal landfill facility as one where 'at least 50 percent of the waste received must be household and/or commercial waste. Household waste is waste that typically comes from residential units. Commercial waste comes from office buildings, restaurants, and other retail and wholesale businesses' (US EPA 1987, p. 15). We shall adopt these definitions for the remainder of this book. Of total waste received at US municipal landfill facilities, 72 percent is household waste, 17 percent is commercial waste, 3 percent is industrial process waste and 6 percent is construction/demolition waste (US EPA 1987, p. G7). Clearly, municipal landfill space is being filled by commercial establishments as well as by households. However, commercial establishments *have* historically been charged user fees. They are charged according to the size of their waste receptacle and the frequency with which it is emptied. Thus, the household sector is where the market for SWS is inefficient and thus where the welfare loss is occurring.

1.2 THE PRIMARY PURPOSE OF THIS BOOK

Our primary goal is to estimate an equation that explains residential demand for SWS. Stated differently, we want an equation that explains the quantity of waste discarded by households. We will attempt to answer two important questions: 'Do households reduce the quantity of waste they discard in response to a user fee for SWS?' and 'How big is the welfare loss to society due to a lack of residential user fees for SWS?' The latter question put positively is, 'How much of a welfare gain will society enjoy if a residential user fee for SWS is imposed?' As a byproduct of our work with residential demand, we will also estimate an equation that explains commercial demand for SWS or the quantity of waste discarded by firms. The latter will reveal how responsive commercial establishments are to user charges for SWS.

1.3 EXISTING RESEARCH ON THE HOUSEHOLD'S RESPONSE TO USER FEES FOR SWS

The original effort by an economist to estimate the response of consumers to volume- or weight-based user charges for SWS is by Wertz (1976). He compares two data points from 1970: the per capita quantity of waste disposed of in a user fee city (San Francisco) and the per capita quantity disposed of, on average, in all other US cities. The latter is taken to represent zero price cities, because most communities in the US do not charge user fees for SWS. Wertz calculates a price elasticity of demand for residential SWS equal to -0.15, meaning a 1 percent increase in a user fee for SWS causes waste to decline by 0.15 percent. Wertz concludes that a user fee has a substantial negative effect on the quantity of waste discarded by households.

Due to its cursory nature, Wertz' empirical analysis can easily be criticized. For example, he does not control for variations across cities in income, population density, weather and other variables that affect the quantity of waste residents discard. We will account for these variables and use more sophisticated econometric techniques and a much larger data set than that used by Wertz.

1.4 FACTORS THAT AFFECT RESIDENTIAL AND COMMERCIAL DEMAND FOR SWS

To determine what variables might affect a household's or firm's demand for SWS we develop two theoretical models: one for the residential and one for the commercial sector. For the former, we develop a utility maximization model in which utility depends positively on the quantity of goods consumed and negatively on the amount of recycling. We include in the household's budget constraint a user charge for SWS. The model suggests that several variables affect residential demand for SWS: the level of household income, the prices of goods consumed, the payment a household can expect for recycling various items, and the user charge for SWS. We hypothesize that residential demand - measured as the daily quantity of waste discarded by a household member - depends on several additional items: the size of the household, the age distribution of the household, the local weather conditions, and the degree of urbanization of the local community.

For the commercial sector, we develop a profit maximization model. We assume that the costs of production increase when a firm recycles but that revenues also go up to reflect the payment that firms receive for used materials. The model suggests that the demand for commercial SWS is a function of the price of the factors of production, the payment the firm receives for recycled materials, the level of employment at the firm and the user charge for commercial SWS. Commercial demand would also seem to depend on the degree of urbanization in the firm's community and on the local weather conditions.

1.5 CHARACTERISTICS OF THE DATA COLLECTED

There is no national agency or association that records the quantities of waste discarded across the US. Thus, to gather data, we called solid waste officials in various communities that charged residents volume-based user fees for SWS. We collected usable data for five such communities. We also collected data for four communities that did not charge user fees for SWS, for comparison with the other data. The data reflect the diversity of solid waste rules and practices that existed in the nine sample communities. An example of this diversity is that open burning is legal in one of the sample communities but illegal in the others, which clearly affects the quantity of yard waste discarded by residents.

To complicate matters further, many of the nine communities did not keep separate records on the quantity of commercial versus residential waste. Most kept track only of the sum of commercial and residential waste. Thus, when we pool the data for all nine communities, we have some dependent variables that reflect the quantity of residential waste only, some that reflect commercial waste only and some that reflect the sum of commercial and residential. We manage to use all of this information by including both the commercial and the residential independent variables on the right hand side of the observations whose dependent variable reflects the sum of commercial and residential waste.

To reduce the chances that the residents represented by our data set are responding to a user fee for SWS by illegally discarding waste, we   only include data from communities that require residents to subscribe to a minimum service level.

1.6 A SUMMARY OF THE EMPIRICAL RESULTS

We estimate the models described above by generalized least squares (GLS). We include a dummy variable for each of the nine communities to capture any unique aspects of that community that might affect the quantity of waste it discards. Thus, the dummy variables should reflect the discrepancies in the data collected across communities - for example, that one community permits open burning while the others do not.

1.61 Results for the Residential Sector

For the residential sector, our model relates the quantity of residential waste discarded per capita per day to the average household income, the regional price being paid for old newspapers, the local temperature and level of precipitation, the average household size, the community's population density, the percent of the population aged 18 to 49 and the user fee charged for residential SWS. We have substituted proxies for several of the variables mentioned above (see Section 1.4). For example, we measure the degree of urbanization as the population density and we measure the payment households receive for recycling as the price being paid for old newspapers. Also, instead of including the price households pay for consumption goods directly in the residential equation, we simply deflate all the monetary variables by the regional consumer price index.

All of the coefficients estimated for the residential equation are significantly different from zero except the price being paid for old newspapers. Table 1.2 gives the sign of each estimated coefficient. Thus, for example, our model suggests that as the average household income of a community increases, the quantity of waste discarded per capita per day increases as well. For a detailed discussion of the reasons the coefficients are either positive or negative see Chapter 6.

The most important coefficient is the negative one estimated for the residential user fee variable. Its actual value is -0.40, which suggests a price elasticity of demand for residential SWS equal to -0.12 at sample means. Let us investigate first whether this coefficient suggests that households are indeed responsive to user charges and second whether the welfare loss triangle pictured in Figure 1.1 is large for a typical US community.

Table 1.2 Signs of the Estimated Coefficients for the Residential Equation

Variable	Sign of Coefficient
User fee for residential SWS	Negative
Average household income	Positive
Local temperature in degrees	Positive
Local precipitation in inches	Positive
Average household size	Negative
Percent of the population aged 18 to 49	Positive
Population density	Positive
Price paid for old newspapers	Negative*

*This coefficient was insignificant at the 5 percent significance level.

To get an idea of how responsive households are to a user fee, let us consider a hypothetical community modelled after a typical US community. Assume that this community discards the average quantity of residential waste discarded by our sample -2.60 pounds per capita per day. Assume further that this community switches from a fee for SWS that is deducted from property taxes to a user charge of $1.31 per 32-gallon container of refuse. The user charge is set to recoup the marginal social cost of SWS in an average US community: $125.00 per ton. This value is suggested by an EPA study that found the cost of disposal alone to be $50.00 per ton in an average community (US EPA 1987, p. G15).[5] Our results suggest that in response to the $1.31 charge, the residential quantity of waste will decrease by 20 percent or by half a pound per person per day. This suggests that the annual quantity of waste discarded per person will go down by nearly 200 pounds. If the hypothetical community has a population of 100,000, then each year 9,600 tons of waste are diverted from the landfill or other disposal facility; with a population of 500,000, 47,800 tons are diverted. Clearly, a user fee policy has the potential to considerably extend the life span of a disposal facility.

Consider now the size of the welfare gain to the hypothetical community. If the community has a population of 100,000, the annual welfare gain would be $600,000. With a population of 500,000, it would grow to $3 million. Thus, we conclude that communities should enjoy a substantial welfare gain when user fees for residential SWS are imposed.

1.62 Results for the Commercial Sector

The commercial equation that we estimate relates the daily quantity of commercial waste per employee to the regional price paid for old corrugated containers (boxes), the community's population density, the local temperature and precipitation and finally, the user charge for commercial SWS. Just as for the residential equation, we have substituted proxies for certain of the variables mentioned in Section 1.4. For example, the price paid for old boxes represents the price paid for all commercially recycled items. Also rather than including the number of employees on the right hand side of the equation we have moved this variable to the left by deflating the quantity of commercial waste by the number of employees. Finally, we have not placed the price paid for factors of production directly into the commercial equation. Instead we deflate all monetary variables by the national producer price index.

Table 1.3 lists the sign of the coefficient estimated for each independent variable in the commercial equation. All coefficients are significant at the 5 percent significance level except the coefficients for the precipitation variable and for the price paid for old corrugated containers. The commercial coefficients suggest, for example, that as the temperature of a community increases the quantity of commercial waste discarded per employee per day also increases. For a discussion of how to interpret the coefficients, see Chapter 6.

As expected, the coefficient for the commercial user fee variable was negative. Its numerical value, -0.23, gives a price elasticity of demand for commercial SWS equal to -0.29 at the sample mean values of the relevant variables. This is substantial and suggests that if commercial fees are raised by 1 percent, the commercial demand for SWS will decline by 0.29 percent. Recall there is no welfare gain to estimate for the commercial sector because historically businesses have always been charged user fees for SWS.

Table 1.3 Signs of the Estimated Coefficients for the Commercial Equation

Variable	Sign of Coefficient
User charge for commercial SWS	Negative
Temperature in degrees	Positive
Precipitation in inches	Negative*
Population density	Negative
Price paid for old corrugated containers	Positive*

*This coefficient was insignificant at the 5 percent significance level.

1.63 The Results as a Forecast Model

An important contribution made by our final estimated demand model is its application as a forecasting tool. The inclusion of variables that reflect the measurable differences across communities makes the model applicable to diverse communities. The residential and commercial equations can be used to predict the future quantities of waste likely to be discarded by a given community. To produce such forecasts, one needs data that represent the independent variables for a previous time period as well as for the forecast period.

1.7 AN OUTLINE OF THIS BOOK

Chapter 2 reviews the existing literature on the response of households to user charges for SWS. In addition, it reviews the findings of others regarding the correlation between waste quantities and such variables as household income and population density.

In the rather technical Chapters 3 and 4 we develop our theoretical and empirical models, respectively. Chapter 3 outlines the utility maximization model of the household and the profit maximization model of the firm. It presents theoretical results that suggest that in response to a positive user fee for SWS, a household will increase its recycling effort and attempt to reduce its consumption of waste intensive goods. Chapter 4 outlines the specifications for our empirical model. It also explains why we use a GLS estimation technique. Both Chapters 3 and 4 can be bypassed or skimmed by readers who are interested primarily in the more practical issues concerning user fees for SWS.

Chapter 5 gives a careful and thorough description of the data set. We provide such a detailed description for several reasons. First, the data set is unique. It includes time series data related to waste quantities for the nine cross-sections or communities, five of which charge a residential user fee for SWS. More important perhaps are the differences in the data available for the nine communities. These differences point out the diversity of laws and conventions that affect the quantities of waste collected by haulers. For example, waste quantities are affected by the availability of recycling centers and by rules regarding the acceptability of construction/demolition debris to a municipal landfill.

The final section of Chapter 5 gives the detailed empirical model. Included is an explanation of the variables in the model and how each is measured.

Chapter 6 gives the empirical results. It lists the coefficients estimated for the independent variables and discusses reasons for the sign of each. Particular attention is given to the coefficient estimated for the residential user fee variable. Its magnitude is discussed and estimates of the welfare gains that should follow imposition of a user fee are given. Although the research is based exclusively on US data, the results hopefully will shed light on the effectiveness of a user fee policy for reducing waste quantities in any community experiencing a shortage of disposal capacity.

Chapter 7 is another rather technical chapter. It consists of three sections, each of which is either a test of the empirical model or a reformulation of it. The most important conclusion reached is that the coefficients in the model should ideally vary from community to community. However when estimating the model separately for each community, so many degrees of freedom are lost that very few coefficients are significant. We conclude that there is a trade-off between the bias introduced by too few observations for each community and that introduced by pooling the data. The latter seems the lesser of the two evils. Chapter 7, like Chapters 3 and 4, can be skipped or skimmed by readers interested only in the practical matters concerning user fees for SWS.

The sole purpose of Chapter 8 is to provide easy-to-follow instructions for forecasting the quantity of waste discarded by a given community. This chapter is designed to stand alone. Thus, readers interested only in forecasting may proceed directly to it. We suggest, however, that such readers eventually peruse Chapter 6 as well to gain insight into the forecasting results.

Chapter 9 presents an overview of the book. It also discusses the implications of our results for policy and gives suggestions for future research.

NOTES

1. The benefit to households of receiving SWS is represented by the demand curve itself. The reason is that if a household is willing to pay a certain dollar value to receive a service, then it must be true that the service benefits the household by that value.
2. See US EPA (1987, p. G3). According to the EPA, to be considered a municipal landfill facility, 'a facility had to receive primarily household and commercial waste, not be classified as a Subtitle C facility, and be at a location where business was conducted or where services or industrial operations were performed by a municipality, corporation, or other public or private entity.' (US EPA, 1987, p. 15).
3. See, for example, Menell (1990, p. 664, n. 32) who cites *Keep America Beautiful, Overview: Solid Waste Disposal Alternatives (1989)* 22. Also see the 1985 private communication between G. Smith of the US EPA and Neal and Schubel as cited by Neal and Schubel (1987, p. 5, n. 11).
4. Dr Ed Repa, the Director of Technical and Research Programs at the National Solid Waste Management Association in Washington, DC suggested during a telephone conversation with the author that siting required an average of five to seven years. Glebs (1988, p. 85) reports that the process of obtaining a permit to open a landfill takes at least two to five years.
5. Please see Chapter 6 for details.

2. A Review of the Literature

2.1 AN OVERVIEW

The existing literature on the demand for SWS has focused on residential demand, almost to the point of excluding commercial demand. The literature consists mostly of empirical work in which residential demand for SWS is related to numerous variables such as a user fee for SWS or household income. With some overlap, existing research papers can be placed into three categories: those investigating volume-based user fees, those investigating service-level-based user fees and those that focus on miscellaneous socioeconomic variables. The papers most relevant to our research are those that have estimated the effect of residential volume-based user fees. By volume-based we mean a user fee that varies according to the size and number of refuse containers. In principle, a user fee could be based on the weight of refuse collected and we will refer to such a fee as weight-based. In practice, however, such a fee is extremely rare and there are no studies of its impact.

Other papers of interest are those that have estimated the effect of service-level-based user fees, such as fees based on the frequency of collection. Finally, numerous papers have studied the effects of various miscellaneous socioeconomic variables such as household income and the size of the household. We will review the most important papers from each of the three categories.

2.2 THE IMPACT OF VOLUME-BASED USER FEES

We have discovered only two studies that attempt to estimate the direct effect of a positive volume-based user fee on residential demand for SWS. The first of these (Wertz 1976) is, in fact, a very cursory empirical investigation. The focus of Wertz's paper is, instead, the development of a formal theoretical model of the demand for SWS. Because Wertz's model is the basis of our own, we will review it in detail. We will review Wertz's empirical findings as well.

15

A second paper that studies the relationship between volume-based user fees and residential demand for SWS is that given by Efaw and Lanen (1979). Because many of their findings are counter intuitive, Efaw and Lanen's research will also be reviewed in detail.

Finally, two additional studies try to measure the impact of volume-based user fees indirectly. McFarland et. al. (1972) and Skumatz (1990) estimate a relationship between the revenues per ton earned by a solid waste collection agency and residential demand for SWS. These researchers considered this relationship because they assumed that the average revenue variable was a proxy for a user fee variable. As we shall discuss further below, replacing a user fee with an average revenue variable is valid only under certain conditions.

Probably the best known economic model of the household's decision regarding how much waste to discard - in other words, a model of the demand for SWS - is that given by Wertz (1976). Wertz analyzes the impact of four variables on the quantity of waste generated by individuals: the price of refuse removal per pound of refuse (or a weight-based user fee); the frequency of collection; the distance an individual must carry refuse to the collection site; and finally, the individual's income level. Wertz assumes that the consumer attempts to maximize utility, a function of the quantities of goods consumed and certain refuse services received. The latter include the frequency and the location of refuse collection. Frequency and location affect utility by affecting, respectively, the accumulation of refuse in one's home and the amount of work required to carry refuse to the collection site.

Wertz's utility maximization is subject to a budget constraint that incorporates the cost of SWS via a positive user fee. From this framework, Wertz derives first order conditions and performs a comparative static analysis. The results give the direction of the relationship between the quantity of waste households generate and a user fee for SWS, the frequency of collection, the distance to the collection site, and the individual's income. We will review only the conclusions regarding the user fee and the individual's income.

Wertz's analysis suggests that the consumer balances the positive utility directly generated by consumption against the disutility associated with the increase in refuse that accompanies consumption. There are two reasons that refuse causes disutility. The first is that on-site accumulation of solid waste offends the senses, takes up space and attracts pests. The second is that the effort spent carrying refuse containers generates negative utility. Wertz concludes that consumption of refuse-generating goods will vary inversely with both the user fee and the absolute value of the disutility of refuse.

Wertz's analysis also suggests that the quantity of waste generated varies positively with income but that this variation, measured as the income elasticity of the demand for SWS, is probably slight.

In sum, Wertz's theoretical model suggests two hypotheses: that the quantity of waste varies negatively with a user fee but positively (and not greatly) with income. Wertz provides empirical support for each of these hypotheses.

To test the latter hypothesis, Wertz collects cross-sectional data for 1970 for 10 suburbs of Detroit. The 10 were selected because they had similar waste collection service characteristics. Using this data and the method of ordinary least squares, Wertz estimates a linear regression model relating annual pounds of refuse collected per capita to annual income per capita. Using the results of this regression, he estimates the income elasticity of waste as 0.279 at the sample means of the variables involved. This suggests that when a consumer's income increases by 1 percent, his waste will increase by 0.279 percent. Repeating this procedure on another cross-sectional sample based on an additional six Detroit suburbs, again, all of which shared collection service characteristics, Wertz estimates the income elasticity as 0.272. The similarity of these estimates suggests to Wertz that the elasticity of waste with respect to income is indeed low and positive. In other words, Wertz concludes that waste should increase slightly with income.

Wertz next attempts to give support for his hypothesis that the quantity of waste discarded varies inversely with a user fee for SWS. For this, he collects only two data points, both from 1970: the quantity of waste discarded per capita in San Francisco, 699 pounds, and the quantity of waste discarded per capita in 'all urban areas' in the US, 937 pounds. San Francisco was selected because residents there pay a volume-based user fee. Wertz compares the per capita quantity of waste in San Francisco to the per capita quantity in all urban areas because most urban areas in the US do *not* charge volume- or weight-based user fees. Thus, Wertz takes the quantity of waste discarded in all urban areas as being discarded by residents for whom the price of disposing of an additional pound of refuse is zero. Based solely on these two data points, Wertz estimates an arc elasticity of waste with respect to the price of SWS equal to -0.15. He concludes that the demand for SWS is substantially responsive to a user fee.

Perhaps the major theoretical oversight of Wertz's paper is his lack of distinction between waste generated and waste discarded. His empirical work considers the latter, while his theory relates to the former. Such a distinction, as we shall see in Chapter 3, becomes important in the presence of recycling. Wertz also ignores that waste consists of different materials such as glass or paper. Again, this oversight is probably because recycling is excluded from his model. Finally, Wertz disregards any waste generated by actions other than consumption such as yard waste. This could be an important oversight, since reports have estimated that yard waste makes up between 6.5 and 17 percent of the household waste stream.[1]

Wertz's empirical analysis is necessarily limited by the meager data on which it is based. He acknowledges that his estimate of the price elasticity of waste, which does not control for variations across urban areas in income or weather conditions or numerous other influential factors, is oversimplified. In general, a more detailed description of Wertz's data would have been enlightening. For instance, it would be useful to know just what is included in the figure he reports as household refuse collected in San Francisco in 1970. San Francisco households consist predominantly of apartment dwellers who are not always subject to user fees for SWS (*1988 Revisions* . . . 1988, pp. 1-3 and 3-75).

Overall, Wertz's paper provides a useful theoretical springboard for studying the effect of price on the quantities of solid waste households discard. The value of his empirical analysis, however, is limited.

Another study of the effect of price on the amount of residential waste set out for collection is that given by Efaw and Lanen (1979). They analyze data for three cities that charge for SWS according to the number of containers set out. The three cities were chosen so that both popular variants of the volume-based user fee concept are represented. The first variant is a container-based system. It requires households to subscribe to SWS for a specified number of containers (of a given size) each week. For the household to change the number of containers subscribed to involves transactions costs. The households are charged according to their subscription level regardless of how full or heavy the containers are. The second user fee variant requires all waste containers for pick-up to be specially marked plastic bags or self-provided containers marked with a sticker or tag. Under this system, the household is charged a price for the plastic bags, stickers or tags, which reflects collection and disposal costs.

Efaw and Lanen present a case study of each city that describes the solid waste system, the data available, and the type of user fee imposed. The conclusion of each case study presents empirical findings.

Perhaps Efaw and Lanen's most original contribution is their discussion of how residential demand for SWS responds to a container-based user fee. Efaw and Lanen suggest that under this system the quantity of waste discarded is a function of the number of containers subscribed to, say C, and C is, in turn, a function of the subscription price per container and disposable income. That is,

$$w = f_1(C, y)$$

$$C = g_1(P_C, y)$$

(2.1)

where w is the quantity of waste, y is disposable income and P_C is the user fee variable which is specified to be the price per container collected. Note that w is measured as the weight and not the volume of refuse. An indirect, negative relationship between the user fee variable, P_C, and the weight of waste discarded is expected; that is,

$$\frac{\partial w}{\partial P_C} = \frac{\partial w}{\partial C}\left(\frac{\partial C}{\partial P_C}\right) < 0$$

(2.2)

or, in elasticity form,

$$\left(\frac{\partial w}{\partial P_C}\right)\frac{P_C}{w} = \left(\frac{\partial w}{\partial C}\right)\left(\frac{C}{w}\right)\left(\frac{\partial C}{\partial P_C}\right)\left(\frac{P_C}{C}\right) < 0.$$

(2.3)

The elasticity form suggests to Efaw and Lanen that the effect of fluctuations in the per-container price of SWS on the weight of waste discarded can be moderated by 'stomping' (i.e., compacting refuse) or by only partially filling containers. To see this, assume that the elasticity of C with respect to P_C is exactly negative one. Then if the elasticity of waste discarded with respect to the number of containers is less than one, there are two implications. When the price per container decreases so that an individual subscribes to an extra container, it may be only partially filled. When the price per container increases so that an individual reduces her subscription level by one container, the contents of the remaining containers may be compacted. This points out that the container-based fee gives households incentive to reduce the volume but not always the weight of their refuse. In other words, the impact of a volume-based user fee on the weight of waste should be smaller than its impact on the volume of waste.

Let us turn now to a consideration of Efaw and Lanen's empirical work. They estimate linear equations that explain the demand for SWS for three user fee cities: Sacramento, California; Grand Rapids, Michigan; and Tacoma, Washington. The particular demand equation estimated varies from city to city for several reasons. Different equations are specified, first, depending on the user fee variant in use in a city and second, according to the characteristics of the data available for a particular city. In addition,

Efaw and Lanen respecify the original versions of their models in response to unexpected or inexplicable results.

For an example of why the demand equations vary from city to city, consider the models for Sacramento and Grand Rapids. Sacramento requires residents to subscribe to a weekly service level and to change that service level imposes transactions costs. Thus the demand for SWS there is hypothesized to depend on the number of containers subscribed to, which in turn depends on the price per container. Compare this model to the one specified for Grand Rapids, which uses a plastic bag system of user fees. The latter model suggests that the demand for SWS depends directly on the price per plastic bag.

Efaw and Lanen collect monthly data for the 1970s for periods of up to four years for each of the three cities. They estimate their models either by ordinary or two-stage least squares. Their final empirical results suggest that household demand for SWS varies positively with income but that the response of demand to a user fee is insignificant. In at least one of the three cities special circumstances might have led to the latter finding. Grand Rapids operates a *non*-mandatory plastic bag system. Households in Grand Rapids have the option of receiving waste collection services from private companies that might charge fees that are completely unrelated to the number of containers set out. Data for the household waste quantities collected by private firms were not available to Efaw and Lanen. Thus, their tonnage data for Grand Rapids represent only the monthly quantities of residential waste hauled by city trucks. This data is clearly less than ideal, since households that discard large quantities of waste will naturally tend to select a private, flat-fee collection service, while those that discard small quantities will most reasonably select the public, plastic bag system. As a result, the price response estimated could understate the response that would exist in the absence of alternative collection services. As the price charged by the city rises, waste per capita does not respond much, because the city's customer base consists of residents who are already low waste generators and thus do not have great latitude to respond to a high price.

Perhaps the major shortcoming of Efaw and Lanen's research is their oversimplification of the demand equation for SWS. The following is typical of the models specified by Efaw and Lanen:

$$\ln(w) = a + b \ln(P_C) + c \ln(y) + d\,(z) \qquad (2.4)$$

where a, b, c and d are parameters, w, P_C and y are as defined above and z is a vector of dummy variables used to represent three of the four calendar seasons. Their goal is to make an empirical contribution. One would therefore expect them to include all independent variables that might reasonably be expected to affect the quantity of waste discarded. A blatant oversight is not considering the prices of goods that produce solid waste. They also fail to include the market price of recycled items that, for obvious reasons, might affect the quantity of waste set out for disposal. As for their data, Efaw and Lanen do not use it as efficiently as possible. They make no effort to pool the data from their user fee cities to estimate a demand equation. Under reasonable conditions, such pooling would lead to more efficient estimators of the parameters involved. Finally, Efaw and Lanen repeatedly change the set of variables in their demand equations without any attempt to theoretically justify or account for these changes. Such reformulations imply pretest problems which lead to biased estimators. Pretest problems also invalidate tests of significance calculated in the usual way.

Efaw and Lanen consistently find that the demand for SWS is inelastic with respect to the price of those services. This is in direct contradiction to Wertz's preliminary finding that the price elasticity is substantial.

McFarland et al. (1972) and Skumatz (1990) have attempted to measure the effect of volume-based user fees by using a solid waste utility's revenues per ton as a proxy for a user fee variable. This approach is less than ideal, since average revenue does not reflect differences in fee structures among communities. A utility will have a positive value for average revenue whether it is charging a volume-based user fee, a service-level-based user fee or a flat fee. Average revenue may appropriately serve as a proxy for a user fee only under certain conditions. In particular, the community being studied should be charging a positive volume-based user fee, and revenues from that user fee should make up a sizable portion of total revenues.

McFarland collects annual cross-sectional data for 1967 through 1968 for 13 cities in California in which SWS are municipally provided. These 13 cities were selected because each was able to provide information on the quantities of household waste collected and in each, 'collection was financed in part or totally by user charges' (McFarland et al. 1972, p. 93). McFarland did not describe the bases of these 'user charges.' However, one gathers from his comments that many of the cities charged service-level-based user fees - such as a fee for collecting refuse from the backyard instead of from the curb - rather than volume-based fees. The former are more common in the US (Stevens 1977, pp. 2-3) but, unlike volume-based fees, do not impose a positive cost per container of SWS on households.

In spite of the prevalence in his sample of a zero unit price[2] for SWS, McFarland estimates a model relating the annual per capita quantity of household waste to the relevant waste utility's revenues per ton and to the community's average per capita income and population density. He takes logs of all variables and then estimates a log linear model by ordinary least squares. This yields a coefficient for the log of average revenue equal to -0.46, which McFarland interprets as the elasticity of demand for SWS with respect to price. This interpretation is quite liberal given the likely absence of volume-based user fees for SWS in his sample of cities.

Why then is the coefficient estimated for the average revenue variable negative and seemingly significantly different from zero? This is probably the result of feedback between average revenue and the quantity of waste discarded. There is some evidence of economies of scale in waste collection.[3] An increase in per capita waste collected could lead to a decrease in collection costs per ton and therefore in the overall costs of SWS per ton of waste discarded. McFarland could erroneously be measuring the extent of this reverse causation.

Skumatz (1990) also proxies a user fee with an average revenue variable.[4] Using annual data for Seattle for 1971 through 1987, Skumatz estimates a model relating the average annual residential pounds discarded per capita to the average household income, the average household size, the average price received for old newspapers, total residential revenues per ton and two dummy variables. The latter account for some outlying values in Seattle's early tonnage data. The first difference of the log of each variable is taken before estimating the model by ordinary least squares. Her results indicate that the coefficient for the average revenue variable is significant and equal to -0.14.

Skumatz interprets the average revenue coefficient as the elasticity of waste discarded with respect to the average rate charged. However, prior to 1981, the price per container for SWS was zero, while average revenue was positive (Skumatz 1990, p. 2). Therefore, in this early period the average revenue proxy may be inappropriate. As for the post-1981 period, the average revenue variable seems to be a good proxy for Seattle's volume-based user fee. The reason is that 75 percent of the utility's revenues are collected from residential customers whose payments to the utility are almost exclusively from volume-based user fees (Skumatz 1990, p. 2). Thus, at least for the post 1981 period, the data considered by Skumatz do not suffer from the same shortcomings as the data considered by McFarland.

In summary, four studies have attempted to measure the impact of volume-based user fees on the demand for SWS. Wertz estimates the elasticity of the demand for SWS with respect to the user fee for SWS as -0.15. Efaw and Lanen find that the user fee charged for SWS is not significantly related to the demand for SWS. McFarland, using average revenue as a proxy for a user fee variable, estimates an elasticity equal to -0.46. Finally, Skumatz, again using average revenue as a proxy for the user fee, estimates an elasticity of -0.14.

2.3 THE IMPACT OF SERVICE-LEVEL-BASED USER FEES AND SERVICE LEVELS

Service-level-based user fees are much more common in the US than are volume-based fees. One study of the effects of these pricing schemes on residential demand for SWS is that given by Stevens (1977). She investigates two service-level-based user fees: a fee that varies according to the frequency of refuse collection and one that varies according to the collection location. A similar study is that of Kemper and Quigley (1976). They attempt to measure the effect of service levels provided free of charge on the demand for SWS. We will present and attempt to reconcile the results of both studies.

Stevens analyzes two cross-sectional data sets, both created from responses to a 1976 survey. The first consists of data related to the weight of household waste in 61 cities throughout the US. The second consists of data related to the 'compacted volume' of household waste in an additional 93 cities. Stevens selected the cities so that in each residents pay service-level-based user fees for SWS.

Stevens' empirical model consists of a system of three linear demand equations. The first represents a household's demand for SWS, hypothesized to be a function of the following six variables: a constant term, the community's average household income, the percent of households subscribing to two rather than one collection(s) per week, the percent of households that subscribe to backyard rather than curb collection, the average number of persons per household and a variable that takes on values between one and three depending on whether an extra fee is charged for yard waste collection. The second equation in Stevens' three-equation system represents the percentage of the population choosing two rather than one collection visit(s) per week. The third equation represents the percentage of the population choosing backyard rather than curb collection.

Stevens estimates the complete system by two-stage least squares and then estimates the individual equations in the system separately by ordinary least squares. She repeats this procedure twice, once using the data that relate to the weight of waste and once using the cubic yard data. Because the results of the different estimation techniques did not differ significantly, Stevens presents and discusses only the coefficients estimated by ordinary least squares.

Stevens' results suggest that only two independent variables affect the quantity of waste discarded by households. These variables are persons per household and frequency of collection, both of which have positive and significant coefficients. Her finding that a larger average household size leads to a significant increase in the dependent variable is expected, because the dependent variable is measured as waste per household.   More interesting is that she finds a significant negative correlation between the demand for SWS and a user fee based on the frequency of service.   This suggests that households might alter their solid waste disposal activities in response to financial incentives.[5]

Kemper and Quigley (1976) also investigate the relationship between the demand for residential SWS and the frequency and location of service. Their approach is different from that of Stevens.   Whereas Stevens measured the impact of service-level-based user fees, Kemper and Quigley measure the impact of service levels provided to residents at no extra charge.   To do this they gather data for 35 municipalities in Connecticut and for 33 collection routes in New Haven, Connecticut.   In all of these regions, households are not offered the option of subscribing to various levels of service.   The waste disposal authorities simply provide one standard service level.   For some of the regions the standard service level includes what we have referred to as 'extras' - two collections per week or backyard pick-up.

Kemper and Quigley pool the data for the 68 Connecticut regions and estimate both a linear and a log linear equation by two-stage least squares. The dependent variable in these equations is the annual tons of waste discarded per household.   The independent variables are the median income per household, the average household size, the annual frequency with which collection visits are made and a dummy variable equal to zero or one depending on whether collection occurs at the curb or in the backyard.[6]

Kemper and Quigley's results indicate that the number of collection visits per year is not significantly related to the annual quantity of waste discarded.   On the other hand, they find that the convenience of having refuse collected from the backyard has a significant and positive effect on the quantity of waste discarded.   These findings are the opposite of those attributed to Stevens.   The primary difference between the two studies seems to be the manner in which waste collection services are offered.   Extra services are offered for a fee in the communities represented by Stevens' data while they are offered for free in the communities represented by Kemper and Quigley's data.

This difference in data leads the researchers to two entirely different definitions of their service level variables.   The variables Stevens uses are the proportions of households subscribing to the greater service levels. Kemper and Quigley define their service level variables as the number of collection visits per year and a dummy variable that indicates whether collection occurs in the backyard or at the curb.   Each researcher's findings can be explained in light of the differences in their data.

Stevens' results reflect feedback among her variables.   In particular, those households that pay extra for backyard collection probably include a disproportionate share of the elderly. One would expect the elderly to produce less waste, on average, than youthful households, given that consumption activities slow during old age.[7] Thus, at a glance, Stevens' results suggest that receiving backyard collection does not encourage waste generation. However, an alternative explanation is that those individuals with low waste generation rates (namely, the elderly) are precisely those who choose to pay extra for backyard collection. Similar feedback might be affecting the correlation that Stevens finds between the frequency of collection and waste quantities. Those households willing to pay extra for a second weekly collection visit are probably big waste generators in the first place. Thus we cannot confidently conclude that the extra visit encourages waste generation. It might be that high waste generation has encouraged subscription to the extra collection visit.

Kemper and Quigley's results do not reflect feedback. In the communities they study, there is no extra charge for backyard collection, so all households take advantage of it. There is no longer any reason to suspect an over-representation of the elderly among such households. Kemper and Quigley's conclusion that the demand for SWS increases in response to the convenience of backyard collection is quite reasonable. Likewise, in Kemper and Quigley's communities there is no charge for an extra weekly collection visit. All households may take advantage of the extra visit so that there should be no over representation of households who generate high waste quantities in the first place. Kemper and Quigley's conclusion that households will not discard more in response to an extra collection visit also appears to be sound.

In sum, both Kemper and Quigley and Stevens conducted interesting and original investigations into how service levels are related to residential demand for SWS. Kemper and Quigley's model comes closer to measuring the isolated impact of an increased service level on the demand for SWS. Their analysis is not affected by the waste disposal characteristics of households that must pay to receive extra services.

2.4 THE IMPACT OF INCOME AND OTHER SOCIOECONOMIC VARIABLES

Many studies have investigated the relationship between the demand for SWS and various socioeconomic variables. Household or per capita income, or some proxy thereof, is virtually always included as an exogenous variable in equations explaining the quantity of waste discarded by households. Other socioeconomic variables, less frequently included, that we will briefly discuss are the average size of the household, the age distribution of the population and the density of the population.

2.41 Income

Virtually all research efforts that investigate the quantity of waste households or commercial establishments set out for disposal include income as a determinant of that quantity. Findings regarding the sign and significance of the income coefficient are far from uniform. We will limit our discussion to the findings of a few of the studies that are particularly insightful. Richardson and Havlicek (1975, 1978)[8] and Rathje and Thompson (1981) study the relationship between household income and the different materials that make up household solid waste. Their results are unique, because they suggest that the relationship between aggregate waste and income might be unpredictable.

Richardson and Havlicek analyze waste collected from select residential routes in Indianapolis, Indiana in 1972. The routes were chosen to reflect a range of average household incomes. Richardson and Havlicek physically separated the household refuse into 11 different materials such as clear glass, aluminum, plastics and grass. They recorded the weight of each of these 11 materials for 24 samples of solid waste that were collected from the various residential routes. Using these data and census data, Richardson and Havlicek estimate 12 equations by ordinary least squares; one related to each of the 11 waste materials plus one related to aggregate waste. They hypothesize that the weight of each waste material collected in a particular region is a linear function of the corresponding average household income, average household size, percentage of the population aged 18 to 51, and percentage of the population that is black. Because data for household income for the specific regions within Indianapolis were not available, they use the average value of a household's property as a proxy for the average income variable.

Richardson and Havlicek's results suggest that there is a significant and positive relationship between household property values and the quantities of green glass, newspapers, grass and aggregate waste discarded. They find a significant and negative relationship between property values and the quantities of aluminum, textiles, plastics and 'garbage/other' discarded. No significant relationship is found between property values and the quantities of clear or brown glass, 'metals other than aluminum,' or 'paper other than newspapers.'

Perhaps Richardson and Havlicek's most unsurprising result is that the quantity of grass discarded increases with property values or income. In general, the value of a property varies positively with the size of the property's yard which, in turn, is positively correlated with the quantity of any yard waste, including grass. A second result that we might have expected is an inverse relationship between property values or income and the quantity of textile wastes. The likelihood that individuals will donate old clothing and other textiles to charitable organizations rather than simply discard them is probably greater for high-income households. In sum, Richardson and Havlicek's study highlights the positive relationship between income and yard waste and the negative relationship between income and textile wastes.

Rathje and Thompson go a step further than Richardson and Havlicek in the analysis of household solid waste. Rathje and Thompson not only separate waste into various material categories such as aluminum and food debris, they also separate it into specific product categories, such as soda cans and milk cartons. Thus, Rathje and Thompson can shed light on why certain materials are correlated with household income by revealing which products seem to vary with income.

Rathje and Thompson study the solid wastes discarded by households in five neighborhoods in Milwaukee during 1978 and 1979. The five neighborhoods were selected so that three income levels would be represented: middle, moderately low and low. Rathje and Thompson use discriminant analysis and simple cross-tabulations to characterize the differences in the types of wastes discarded by the three income groups. Their most important finding is that packaging waste, overall, was higher for the low-income neighborhoods than for the others. Packaging in the form of paper, aluminum, other metals, and plastic was more abundant in the low-income neighborhoods.

Rathje and Thompson attempt to explain this finding by analyzing the products associated with the solid waste. They conclude that packaging waste was higher among the low-income families for two reasons. First, the low-income families bought more 'preprepared foods and beverages;' in particular, more 'highly processed beverages' such as soda and more grain products and chips. Second, the low-income families purchased small quantities of items; that is, they bought large numbers of small packages of food and other items. Rathje and Thompson suggest that the latter was due to a binding budget constraint and a lack of storage space. They conclude that the low income groups' 'purchasing patterns are not frivolous . . . [but] seem to be the result of (1) the use of relatively inexpensive high-bulk foods, such as breads, chips and pastries . . . and (2) limited cash flow, limited storage space and small family sizes' (Rathje and Thompson 1981, p. 64).

An implication of both the Rathje and Thompson and the Richardson and Havlicek studies is that there is no simple association between income and aggregate household solid waste. There are, however, correlations between income and more narrowly defined waste materials. The quantities of some types of waste increase with income (yard waste and newspapers), while the quantities of other types of waste decrease with income (packaging and textiles). The majority of studies that we have perused find a positive association between aggregate waste and income. This implies that the extra newspaper and yard wastes discarded by high-income families frequently outweigh the extra packaging and textile wastes discarded by low-income families.

2.42 Average Household Size

Many studies have estimated the relationship between the quantity of residential waste and the average household size of a community; i.e., the average number of members of a household. Often, these studies measure the dependent variable as the quantity of waste per household. When this is the case the coefficient estimated for household size is usually positive, as expected, but not significant.

Kemper and Quigley's (1976) intracity study of New Haven, Connecticut provides an approach to estimating the economies of scale within the household. They replace the dependent variable, waste per household, with waste per capita. If there are economies of scale within the household, waste per capita should decrease as the average household size in a community goes up. There are several reasons to expect scale economies. One is that as families grow, larger units of goods are purchased. Thus, the increase in packaging waste associated with an additional family member is quite small. Another is that large families are a natural setting for the handing down of clothing, toys and other such goods from one family member to another. Finally, large families share certain other goods like newspapers and yard space. Despite expectations, the coefficient Kemper and Quigley estimate for household size is positive and insignificant. This suggests that there are no economies of scale within the household.

2.43 The Age Distribution of the Population

A few studies have estimated the relationship between the age distribution of a population and the quantity of waste that the average household discards. Richardson and Havlicek explain why one might expect a correlation between the age distribution and waste quantities:

> The individual's consumption bundle changes in quantity and composition over his lifetime . . . and this may affect the quantity and composition of waste generated. Total waste generated . . . is hypothesized to be low in the early stages of life, increase through middle age, then decline in old age (Richardson and Havlicek 1978, p. 105).

Richardson and Havlicek's (1975, 1978) findings do not fully confirm their expectations. Their research investigates aggregate quantities of waste as well as the materials that make up solid waste. Models that relate to the former sometimes yield coefficients for the percent of the population aged 18 to 61 that are significant and positive. Models that relate to the latter always yield insignificant 'age' coefficients. Clearly, to draw any definite conclusions we need more evidence regarding the correlation between the age distribution of the population and the demand for SWS.

2.44 Population Density

Cargo (1978) and Kemper and Quigley (1976) have tested whether a community's demand for SWS depends systematically on the population density of the community. Presumably these researchers consider their population density variables as measures of the extent of urbanization. Communities' consumption patterns and therefore the quantity of waste they discard might vary according to how urban they are. Whether waste quantities should increase or decrease is not suggested by either researcher.

Cargo estimates the relationship between population density and three different dependent variables: the quantities of residential solid waste, of commercial solid waste, and of mixed (residential and commercial) solid waste. In all three equations the coefficient for population density is positive but not significant.

For their intracity study of New Haven, Connecticut, Kemper and Quigley relate population density to the quantity of residential waste per household and also per capita. The coefficient of the population density variable was significant only in the equation corresponding to the per capita formulation. In that equation the coefficient was negative.

The relationship between urbanization and the demand for SWS is probably complicated. On one hand, as urbanization increases we expect smaller yard sizes and less yard waste. This might have been the explanation for Kemper and Quigley's negative coefficient. On the other hand, as populations become more urban, households have smaller storage spaces. This might lead urban households to discard and later replace items, such as gift boxes or used car parts, that a rural household would simply save and re-use. It could also lead households to purchase goods in small quantities, which leads to a high proportion of packaging to product. In sum, changes in population density suggest that particular refuse items should become more abundant and others less abundant. As with the household income coefficient, the coefficient for population density may frequently be estimated as insignificant. The reason is that the changes associated with urbanization in the various materials that make up solid waste might counteract one another.

2.45 Conclusions Regarding the Socioeconomic Variables

The literature concerned with the impact of various socioeconomic variables on the quantity of waste households discard suggests several conclusions. Income, though not always correlated with the quantity of aggregate waste disposed, does seem to vary systematically with the various materials and products that make up household waste, such as yard waste and packaging waste. The number of members of a household is positively correlated with the aggregate quantity of waste discarded by the household. However, the hypothesis that there are economies of scale within a household is not supported. Estimates of the relationship between the age distribution of the population and the quantity of waste that households discard are contradictory, so we do not draw conclusions. Finally, the aggregate quantity of waste discarded by households is not clearly related to the population density of a region, a proxy for the extent of urbanization. As was the case for income, we expect some material and product categories of waste to be inversely correlated with the population density, while we expect others to be positively correlated with it.

NOTES

1. University of Arizona's Garbage Project (Church 1988, p. 82) reports yard waste as making up 17.1 percent of the weight of refuse discarded by the average Tucson household. However, their estimate of the percentage of yard waste declines dramatically to 6.5 percent based on household refuse samples gathered from three US cities: Tucson, Milwaukee and Marin County, California (Rathje, Reilly and Hughes 1985, p. 30).
2. By units of SWS, we mean the pounds or cubic feet of refuse to which SWS are applied.

3. See, for example, McFarland, et. al. (1972, pp. 59-61) or see the US Environmental Protection Agency memorandum, dated July 1977, from Barbara Stevens to Fred L. Smith as cited by the Organization for Economic Co-operation and Development (1981, p. 47, n. 15).
4. The following discussion draws from Skumatz' 1990 paper and from a supplemental document authored by J. Bagby (1988).
5. We conclude that Stevens' results are important despite several econometric errors in her empirical procedures. For a more complete review of her econometric work see Jenkins (1991) pp. 37-42. Please note that Stevens' paper is a draft that, to our knowledge, was never subsequently revised.
6. Kemper and Quigley provide no details regarding additional exogenous variables on which the frequency of collection might depend. Presumably, they have not expanded their set of exogenous variables when estimating by two-stage least squares.
7. This is a typical assumption of Modigliani's life cycle theory of consumption. We discuss below the findings of other researchers regarding the impact of the age distribution of the population on the quantity of waste discarded by households.
8. The following discussion refers to both papers.

3. Models of the Household and the Firm

3.1 INTRODUCTION

Households produce solid waste as a byproduct of the consumption of commodities. They determine the quantities of waste to recycle and discard by allocating their time among competing activities. Their demand for SWS should, therefore, be related to their maximization of utility subject to the relevant budget constraint. Firms, on the other hand, generate solid waste as an inadvertent result of the production or distribution of goods and services. They determine the quantities of waste to recycle and discard by allocating labor between production and recycling activities. Firms settle on their demand for SWS by maximizing profits subject to output being produced according to the firm's technology or production function.

3.2 A MODEL TO EXPLAIN THE HOUSEHOLD'S DECISIONS REGARDING SOLID WASTE

The household's decision regarding the quantity of waste to discard or recycle is centered around the issue of time. Recycling generally requires more of the household's time than that required to discard waste. Because of the central role of time, we have chosen to analyze the household's utility maximization problem within the framework of the well known household production model.

Assume household utility is a function of the quantities of market goods the household consumes, $X_1,...,X_n$, and the quantity of time the household spends enjoying leisure activities, t_L. More specifically, assume

$$U(X_1,...,X_n, t_L) ;$$

$$U_i > 0, \ i = 1,...,n; \qquad U_{t_L} > 0 ;$$

(3.1)

32

where $U(.)$ represents the utility function and subscripted U's denote partial derivatives of utility. Following Wertz (1976), assume that each unit of the ith good consumed has some fixed component of waste, w_i. For ease of presentation, assume that w_i is a measure of the weight rather than the volume of waste.[1] Assume further that the fixed component of waste, w_i, can be broken down into J types of waste; e.g., the waste component associated with a bag of bread consists of both plastic and food waste. Let w_{ij} be the weight of the jth type of waste associated with one unit of good i. Then

$$w_i = \sum_{j=1}^{J} w_{ij} \; ; \quad i = 1,...,n \; . \tag{3.2}$$

Besides waste generated by market goods, there is also household waste that comes from nonmarket goods. The major component of the latter is yard waste which makes up between 6.5 and 17 percent of the household waste stream.[2] Let Z be the weight of waste created by nonmarket goods and let W be the total weight of all refuse generated by the household. Then, since W is the sum of the weights of all waste components associated with all market and nonmarket goods,

$$W = \sum_{i=1}^{n} \sum_{j=1}^{J} w_{ij} X_i + Z \; . \tag{3.3}$$

To reiterate, W, Z, w_i, and w_{ij} may easily be redefined so that they reflect the volume of waste rather than its weight.

Assume that of the J types of waste, only a subset is recyclable. In actuality, this subset is probably quite large although there are always a few types of waste that cannot be recycled. Included in the latter are some forms of plastic and waste products that are composed of mixed materials, e.g., glossy magazines whose pages consist of paper coated with wax.

Assume further that households receive an actual monetary payment for only a subset of the goods that are recyclable. Many neighborhoods, for example, have recycling centers that pay households for empty aluminum cans. On the other hand, yard and food wastes are rarely purchased from households even though they are quite recyclable. Such waste readily decomposes into fertile topsoil if community or backyard compost piles are maintained.

Thus, from the J types of waste, three categories are relevant to the consumer's utility maximization decision: recyclable waste for which the household receives payment; recyclable waste for which no payment is received; and nonrecyclable waste. Let the first J_1 types of waste be waste that is recyclable and for which the household receives a monetary payment. Let W_{R+M} be the total weight of such waste. Then,

$$W_{R+M} = \sum_{i=1}^{n} \sum_{j=1}^{J_1} w_{ij} X_i \ . \tag{3.4}$$

Let the total weight of waste that is recyclable but for which the household receives no payment be denoted W_R. Assuming that the next J_2 types of waste fall into this category,

$$W_R = \sum_{i=1}^{n} \sum_{j=J_1+1}^{J_1+J_2} w_{ij} X_i + Z \ . \tag{3.5}$$

Waste created by nonmarket goods, Z, is included here since it consists primarily of yard waste which, as mentioned, is recyclable but for which payment is rarely received. Finally, let the total weight of waste that is not recyclable be denoted W_{NR}. Assuming that the last J_3 types of waste fall into this category,

$$W_{NR} = \sum_{i=1}^{n} \sum_{j=J_1+J_2+1}^{J_1+J_2+J_3} w_{ij} X_i \ . \tag{3.6}$$

Households, in general, do not recycle 100 percent of their recyclable waste. The reason is that recycling requires an investment of time. Such an investment has an opportunity cost. Any time the household spends recycling cannot be spent working for wages or enjoying leisure activities. Define R_1 as the quantity of W_{R+M} that is recycled. Likewise define R_2 as the quantity of W_R recycled. Clearly R_1 and R_2 are positive functions of the time that the household invests in recycling; that is,

$$R_1 = R_1(t_1), \quad R_2 = R_2(t_2) \ ;$$

$$\frac{dR_1}{dt_1} > 0, \quad \frac{dR_2}{dt_2} > 0 \ ; \tag{3.7}$$

where
t_1 = the number of hours per period that the household spends recycling W_{R+M} and

t_2 = the number of hours per period that the household spends recycling
       W_R.

It seems likely that items easiest for the household to recycle are recycled
first. If so, then as the quantities of waste recycled, R_1 and R_2, increase, the
time required to recycle one more unit of waste also increases. In other
words, the second derivatives of the quantities of waste recycled with
respect to the time spent recycling are negative,

$$\frac{d^2R_1}{dt_1^2} < 0; \quad \frac{d^2R_2}{dt_2^2} < 0 . \tag{3.8}$$

As well as being intuitively sound, (3.8) is sufficient to ensure satisfaction
of the second order conditions of the household's utility maximization
problem, which is given by (3.10) and (3.11).

Finally, note that R_1 and R_2 have upper bounds. The household cannot
recycle more waste than it has generated. In particular,

$$R_1 \leq \sum_{i=1}^{n} \sum_{j=1}^{J_1} w_{ij} X_i; \quad R_2 \leq \sum_{i=1}^{n} \sum_{j=J_1+1}^{J_1+J_2} w_{ij} X_i + Z . \tag{3.9}$$

Let
P_i = the price of good i;
P_w = the price per unit weight of residential SWS (a weight-based user
       fee);
P_R = the payment received per unit weight for recycling W_{R+M};
V = nonwage household income;
s = the hourly wage rate earned by the household;
T = the total time endowment of the household in hours (e.g., 24 hours
       in a day); and
t_s = the number of hours that the household spends working for wages
       per period.

The household's problem is to maximize

$$U(X_1,...,X_n, t_L) \tag{3.10}$$

subject to:

$$\left. \begin{aligned} s\,t_s + V + P_R\,R_1 &= \sum_{i=1}^{n} P_i\,X_i + P_w(W - R_1 - R_2)\ ; \\[1ex] T &= t_s + t_L + t_1 + t_2\ ; \\[1ex] R_1 &= R_1(t_1)\ ; \\[1ex] R_2 &= R_2(t_2)\ ; \\[1ex] W &= \sum_{i=1}^{n}\sum_{j=1}^{J} w_{ij}\,X_i + Z\ ; \\[1ex] R_1 &\le \sum_{i=1}^{n}\sum_{j=1}^{J_1} w_{ij}\,X_i\ ; \\[1ex] R_2 &\le \sum_{i=1}^{n}\sum_{j=J_1+1}^{J_1+J_2} w_{ij}\,X_i + Z\ . \end{aligned} \right\} \qquad (3.11)$$

We assume that the parameters of the model are such that the household never recycles all of W_{R+M} or all of W_R; that is, the solutions for R_1 and R_2 are interior ones. Thus, the two inequality constraints in (3.11) do not affect the household's decisions and can be dropped from consideration. Collapsing the five equality constraints in (3.11) into one, we may rewrite the maximization problem in Lagrangean form:

$$\begin{aligned} L = {}& U(X_1,...,X_n,\,t_L) + \lambda\{s\,T + V + P_R R_1(t_1) \\[1ex] & - \sum_{i=1}^{n} P_i\,X_i - s(t_L + t_1 + t_2) \\[1ex] & - P_w[\sum_{i=1}^{n}\sum_{j=1}^{J} w_{ij}\,X_i + Z - R_1(t_1) - R_2(t_2)]\} \ . \end{aligned} \qquad (3.12)$$

Differentiating this Lagrangean with respect to $X_1,...,X_n$, t_L, t_1, t_2 and λ gives the following first order conditions.

$$U_i + \lambda[- P_i - P_W \sum_{j=1}^{J} w_{ij}] = 0\ ; \qquad (3.13)$$

$$i = 1,...,n\ .$$

$$U_{t_L} - \lambda s = 0 \ . \tag{3.14}$$

$$\lambda[P_R \frac{dR_1}{dt_1} - s + P_W \frac{dR_1}{dt_1}] = 0 \ . \tag{3.15}$$

$$\lambda[- s + P_W \frac{dR_2}{dt_2}] = 0 \ . \tag{3.16}$$

$$s\,T + V + P_R\,R_1(t_1) - \sum_{i=1}^{n} P_i\,X_i - s(t_L + t_1 + t_2) \tag{3.17}$$

$$- P_W[\sum_{i=1}^{n}\sum_{j=1}^{J} w_{ij}\,X_i + Z - R_1(t_1) - R_2(t_2)] = 0 \ .$$

Let us consider equation (3.13). By placing all terms multiplied by λ on the right hand side of the equality we obtain

$$U_i = \lambda(P_i + P_W \sum_{j=1}^{J} w_{ij}) \ ; \tag{3.18}$$

$$i = 1,...,n \ .$$

Forming a ratio of (3.18) for goods i and h yields

$$\frac{U_i}{U_h} = \frac{P_i + P_W \sum_{j=1}^{J} w_{ij}}{P_h + P_W \sum_{j=1}^{J} w_{hj}} \ . \tag{3.19}$$

The right hand side of (3.19) reflects the marginal cost of good i over the marginal cost of good h. Thus, condition (3.19) suggests that the household will set the marginal rate of substitution between two goods equal to the ratio of the goods' marginal costs. Note that the marginal cost of each good includes both the market price of the good and the cost of disposing of the waste generated by consuming one unit of the good.

Condition (3.19) helps explain the household's response to an increase in the waste associated with a good. Recall that P_w is the price per unit weight of SWS; that is, P_w is a weight-based user fee for SWS. Assume that this user fee is nonzero. Then, everything else equal, the household will respond to an increase in the quantity of waste associated with good *i*, perhaps due to a more bulky package, by decreasing consumption of good *i* relative to good *h*. The reason for this is that the value of the right hand side of (3.19) will rise as the quantity of waste associated with good *i* increases. Hence, the value of the left hand side must also rise. This suggests that the marginal rate of substitution between goods *i* and *h* increases. We can conclude that consumption of the good whose waste has increased declines relative to consumption of the other good.

Of course, to reach this conclusion we assumed 'everything else remains equal.' By this assumption we intend to imply that the additional waste that accompanies good *i* does not affect the consumer's relative valuation of good *i*; that is, the additional waste does not twist the indifference curve that relates good *i* to good *h*. This rules out several very real possibilities. One is that an increase in packaging serves as an effective advertisement. Another is that the additional packaging improves the convenience of the product.

Condition (3.19) suggests a household will reduce consumption of a good when the quantity of waste associated with it increases whether the additional waste is recyclable or not. At first glance this seems unreasonable. If waste is recyclable, the household can potentially avoid any disposal costs. However, as analysis of the remaining first order conditions will elucidate, avoiding disposal costs entails another cost, the cost of the household's time. We shall show below that the full cost of the marginal unit of a good is always equivalent to its price plus the cost of disposing of its waste.

Let us restate the implications of equation (3.19). When paying a nonzero user fee for SWS, households will increase their consumption of a good in response to a decline in the waste associated with the good. This suggests a rather important conclusion. The presence of a nonzero user fee gives incentive to manufacturers to decrease the waste associated with their products.

First order condition (3.14) can be expressed as

$$U_{t_L} = \lambda \, s \, . \tag{3.20}$$

Interpreting λ as the marginal utility of income[3] and *s* as the marginal income of time spent working, equation (3.20) suggests that the optimizing household sets the marginal utility of time spent at leisure equal to the marginal utility of time spent working.

First order conditions (3.15) and (3.16) relate to the household's decision regarding the time to spend recycling. Let us rewrite equation (3.15), dividing all terms by λ and placing the wage rate, s, on the right hand side of the equality.

$$\frac{dR_1}{dt_1}(P_R + P_W) = s .$$ (3.21)

Likewise, let us rewrite equation (3.16):

$$\frac{dR_2}{dt_2} P_W = s .$$ (3.22)

Equations (3.21) and (3.22) suggest that the optimizing household recycles until the value of the marginal product of time spent recycling equals the wage rate. The left hand side of (3.21) reflects the value of the marginal product of time spent recycling marketable waste. This time is valuable because it saves the household disposal costs plus it earns the household the market value of the recycled items. The left hand side of (3.22) reflects the value of the marginal product of time spent recycling nonmarketable waste. The household receives no payment for recycling this waste; the time is valuable, however, because it saves the household disposal costs.

Consider equations (3.21) and (3.22) with all price variables on the right hand sides of the equalities:

$$\frac{dR_1}{dt_1} = \frac{s}{P_R + P_W} ;$$ (3.23)

$$\frac{dR_2}{dt_2} = \frac{s}{P_W} .$$ (3.24)

Recall we have assumed that any time the household spends recycling will increase the quantity of waste recycled but at a decreasing rate. Thus, equations (3.23) and (3.24) suggest that, everything else equal, the number of hours spent recycling varies directly with the user fee for SWS.[4] The optimizing household will respond to an increased user fee, P_w, by reducing the marginal products of both t_1 and t_2. These reductions can only be brought about by increasing the hours spent recycling both marketable and nonmarketable waste.

Assume for the moment, in addition to the assumption of declining marginal productivity of the time spent recycling, that the relationship between R_1 and t_1 is identical to the one between R_2 and t_2 - that if t_1 and

t_2 are equal then the quantity of marketable waste recycled equals the quantity of nonmarketable waste recycled. A comparison of equations (3.23) and (3.24) suggests that under these special conditions the household will recycle a greater quantity of marketable waste than nonmarketable waste; that is,

$$R_1 > R_2 \ . \tag{3.25}$$

The reason for this is that the marginal product of the time spent recycling marketable waste (the left hand side of equation (3.23)) is chosen so that it equals the wage rate divided by the sum of the price of SWS and the price received for recyclables. The marginal product of the time spent recycling nonmarketable waste is, however, equated to the wage rate divided by the price of SWS only (see equation (3.24)). Thus, under these special circumstances, the marginal product of t_1 is less than the marginal product of t_2 which implies that t_1 is greater than t_2.

Given the household's decision rules governing selection of t_1 and t_2, we can better understand the full marginal cost of a market good, say X_i. Recall that equation (3.19) suggests that the marginal cost of a good includes the price of the good plus the cost of disposing of the waste generated by consuming it. Equation (3.19) is relevant for all goods whether they are associated with marketable waste, recyclable but nonmarketable waste or nonrecyclable waste. This might seem counterintuitive since the household has the option of avoiding disposal costs when faced with recyclable waste. The household can not only avoid disposal costs but also receive a monetary payment if it recycles marketable waste. Why then does the full marginal cost of all goods, regardless of the disposition of the waste they generate, include the cost of disposing of their consequent waste? Consider a good, say X_h, whose waste is recyclable but not marketable; that is,

$$\sum_{j=1}^{J_1} w_{hj} = 0, \quad \sum_{j=J_1+J_2+1}^{J_1+J_2+J_3} w_{hj} = 0 \ . \tag{3.26}$$

Assume that the household recycles 100 percent of the waste associated with the marginal unit of X_h. The number of hours required to accomplish such recycling can be approximated by dividing the quantity of waste to be recycled, w_h, by the marginal product of the last hour spent recycling:

$$\frac{w_h}{\dfrac{dR_2}{dt_2}} \cdot \tag{3.27}$$

The cost to the household of this recycling is the opportunity cost of the time invested to accomplish it; that is,

$$s\,\frac{w_h}{\dfrac{dR_2}{dt_2}} \cdot \tag{3.28}$$

We can now see that the full marginal cost of good h to the household is its price plus the cost of the time required to recycle it:

$$P_h + s\,\frac{w_h}{\dfrac{dR_2}{dt_2}} \cdot \tag{3.29}$$

Via equation (3.24), we can replace dR_2/dt_2 in (3.29) with s/P_w. Doing so reveals that the full marginal cost of good h is equivalent to

$$P_h + P_W\,W_h \cdot \tag{3.30}$$

This is precisely the marginal cost that we originally found questionable; that is, the marginal cost suggested by the first order condition for good h.

We can also show, following similar steps, that the marginal cost of a good whose waste is marketable can accurately be reflected by the price of the good plus the cost of disposing of the good's waste. This result may be shown via equation (3.23).

We conclude that there are two important consequences of a positive user fee for SWS. The first is that increases in the quantity of waste associated with a good lead to increases in the good's marginal cost, regardless of the quality of the waste. The other is that the household will increase the time spent recycling in response to a positive user fee for SWS; that is, there is a positive relationship between the quantity of waste recycled and a user fee.

Assume now that the $n + 4$ first order conditions (3.13) through (3.17) can be solved for $X_1,...,X_n$, t_1 and t_2. These solutions will be functions of the parameters of the system. Denote these solutions as

$$X_i = G_i(s, V, P_1,...,P_n, P_R, P_W, w_{11},...,w_{nJ}, Z),$$

$$i = 1,...,n \ ;$$

$$\tag{3.31}$$

$$t_1 = G_{t_1}(s, V, P_1,...,P_n, P_R, P_W, w_{11},...,w_{nJ}, Z) \ ;$$

$$t_2 = G_{t_2}(s, V, P_1,...,P_n, P_R, P_W, w_{11},...,w_{nJ}, Z) \ .$$

Expressing the demand for SWS, W^d, in terms of the weight of total discarded waste, and using evident notation based on (3.31):

$$W^d = W - R_1 - R_2$$

$$= \sum_{i=1}^{n} \sum_{j=1}^{J} w_{ij} \ G_i(.) + Z - R_1[G_{t_1}(.)] - R_2[G_{t_2}(.)] \tag{3.32}$$

$$= F(s, V, P_1,...,P_n, P_R, P_W, w_{11},...,w_{nJ}, Z)$$

where $F(.)$ is a function of the indicated arguments. Thus, the demand for SWS is a function of the wage rate earned by the household, the household's nonwage income, the prices of goods, the price received for recyclables, the price of residential SWS or the user fee, the waste components of all goods consumed and the quantity of waste generated by nonmarket goods. Estimable forms of equation (3.32) are discussed in Chapter 4. At this point we only note that (3.32) indicates the arguments of the residential or household demand function for SWS.

3.3 A MODEL TO EXPLAIN THE FIRM'S DECISIONS REGARDING SOLID WASTE

Commercial sector waste is the result of the production and distribution activities of firms. Just as the household produces waste from goods that generate utility, the firm produces waste from inputs that, when combined in certain ways and according to certain processes, produce output. SWS enter the firm's decision-making process as services whose costs, along with the costs of inputs, must be subtracted from revenues. Recycling, on the other hand, affects the firm by drawing labor away from production activities. Recycling can also affect direct sales if the recycled items are marketable.

As the following discussion reveals, our analysis of the firm closely parallels our analysis of the household. We have ensured such a close correspondence by modelling the household's utility maximization problem within a household production framework. Thus, the conditions suggested by the firm's first order conditions for profit maximization are analogous to those that emerged from our analysis of the household's problem.

Assume that the firm chooses the level of output that maximizes profits. This level of output must be produced according to the relevant production function. Let

Q = the quantity of output produced;
I_i = the quantity of nonlabor input i; and
L_p = the number of labor hours used in production.

Assume that

$$Q = f(I_1,...,I_M, L_p) ;$$

$$f_i > 0, \quad i = 1,...,M; \quad f_{L_p} > 0 ;$$

(3.33)

where $f(.)$ is the production function and subscripted f's denote partial derivatives of the production function. We have assumed in (3.33) that output is a function of the quantities of nonlabor inputs used, $I_1,...,I_m$, and the quantity of labor used in production, L_p.

Because the firm generates waste as a byproduct of the process of production or distribution, we assume that the weight of commercial waste, W^c, is a function of a subset of the nonlabor inputs used in the production process:

$$W^C = W^C(I_1,...,I_{M-j}) .$$

(3.34)

This seems appropriate since commercial waste includes such items as used up inputs, the packaging that accompanies inputs, and the unusable components of inputs. Equation (3.34) is in line with assumptions made by economists concerned with the air emissions of firms. In that literature, emissions are frequently written as a function of a subset of the inputs used in the production of output.[5] Residuals - in the present context, solid waste - are a function of only those nonlabor inputs that either contribute to the creation of waste or lead to its reduction. Presumably, there may exist some production inputs that do not affect the quantity of waste at all.

Because W^c in (3.34) represents the quantity of waste *generated*, it is not a function of labor. We assume that labor affects waste by affecting the quantity *discarded* through recycling efforts. Thus, labor's important role is acknowledged by separating labor hours into those devoted to production activities versus those devoted to recycling activities.

Rather than express the quantity of commercial waste as a general function of a subset of nonlabor inputs, for the sake of simplicity, we adopt the stronger assumption that it is a linear combination of such inputs.

$$W^C = \sum_{i=1}^{M} w_i \, I_i \qquad (3.35)$$

where w_i is the weight of waste associated with one unit of the ith input and w_i can be zero or negative as well as positive. Admittedly, assuming that w_i is fixed across firms is a simplification, perhaps more so than the analogous assumption for the household. The use of goods across consumers and, hence, the waste generated by each good is probably more uniform than the use of inputs across producers and the waste associated with each input. This problem can be overcome by defining an input not just according to its physical structure but according to the use to which it is put as well. With this narrow definition of 'input,' the assumption that a fixed portion of each input eventually becomes waste is not so unrealistic.

For the firm, similar to the household, waste can be broken down into K types. The waste component associated with each input can be written as the sum of these K types of waste. That is,

$$w_i = \sum_{k=1}^{K} w_{ik}, \quad i = 1,...,M \qquad (3.36)$$

where w_{ik} is the weight of the kth type of waste generated by one unit of input i. These K types can, in turn, be grouped into three categories. Some types of commercial waste can be sold and are thus marketable. Label this waste W^C_{R+M}. If the firm re-uses certain types of its waste as inputs into its own production process, this is equivalent to the waste being marketable. The dollars saved by not purchasing virgin inputs are equivalent to a price received for the waste. Thus, waste re-used within the firm can be grouped with marketable waste. Assuming that the first K_1 types of waste are marketable, then,

$$W^C_{R+M} = \sum_{i=1}^{M} \sum_{k=1}^{K_1} w_{ik} \, I_i \, . \qquad (3.37)$$

The firm typically will not recycle all of this waste. The quantity that it does recycle, say R_1^C, depends positively on the number of labor hours devoted to recycling marketable waste, L_1. Specifically,

$$R_1^C = R_1^C(L_1); \quad \frac{dR_1^C}{dL_1} > 0 . \tag{3.38}$$

Assuming that the easiest recycling is done first, as more labor is devoted to recycling the productivity of labor falls off; that is,

$$\frac{d^2 R_1^C}{dL_1^2} < 0 . \tag{3.39}$$

The assumption represented by (3.39) is not only reasonable but is sufficient to satisfy the second order conditions of the firm's profit maximization problem given by (3.45) and (3.46).

Of course, the quantity recycled can never be greater than the total quantity of marketable waste generated by the firm; that is,

$$R_1^C \leq \sum_{i=1}^{M} \sum_{k=1}^{K_1} w_{ik} I_i . \tag{3.40}$$

Firms have other types of waste that are recyclable but for which payment may not be received; e.g., corrugated containers (boxes). The avoidance of disposal costs gives the firm incentive to recycle this waste. Assume that the next K_2 types of waste fall into this category, W_R^C.

$$W_R^C = \sum_{i=1}^{M} \sum_{k=K_1+1}^{K_1+K_2} w_{ik} I_i . \tag{3.41}$$

The quantity of nonmarketable waste recycled by the firm is R_2^C. This quantity varies directly with the number of labor hours spent recycling nonmarketable waste, L_2, and can never be greater than W_R^C, the total quantity of recyclable but nonmarketable waste generated by the firm. In particular,

$$R_2^C = R_2^C(L_2); \quad \frac{dR_2^C}{dL_2} > 0; \quad R_2^C \leq \sum_{i=1}^{M} \sum_{k=K_1+1}^{K_1+K_2} w_{ik} I_i . \quad (3.42)$$

Additionally, we assume that as more and more labor time is spent recycling nonmarketable waste, the productivity of that time diminishes:

$$\frac{d^2 R_2^C}{dL_2^2} < 0 . \quad (3.43)$$

The latter assumption will suffice for the second order conditions of the firm's profit maximization problem to be satisfied.

Finally, firms have some types of waste that are virtually nonrecyclable. There are always some solid waste residues which must be discarded. Denote nonrecyclable wastes as W_{NR}^C, and assume that the last K_3 types of waste fall into this category; then,

$$W_{NR}^C = \sum_{i=1}^{M} \sum_{k=K_1+K_2+1}^{K_1+K_2+K_3} w_{ik} I_i . \quad (3.44)$$

Let

P_q = the price per unit of output, Q;
r_i = the price per unit of input i;
s = the hourly wage rate;
L = the total quantity of labor used by the firm, in hours;
P_W^C = the price paid by the firm per unit weight of SWS (a commercial weight-based user fee);
P_R^C = the price received by the firm per unit weight for recyclables.

The firm wishes to maximize

$$P_q Q - \sum_{i=1}^{M} r_i I_i - s L \\ - P_W^C (W^C - R_1^C - R_2^C) + P_R R_1^C \quad (3.45)$$

with respect to $Q, I_1, ..., I_M, L_P, L_1,$ and L_2 and subject to

$$Q = f(I_1,...,I_M, L_P) \; ;$$

$$L = L_P + L_1 + L_2 \; ;$$

$$W^C = \sum_{i=1}^{M} \sum_{k=1}^{K} w_{ik} I_i \; ;$$

$$R_1^C = R_1^C(L_1) \; ;$$

$$R_2^C = R_2^C(L_2) \; ;$$

$$R_1^C \leq \sum_{i=1}^{M} \sum_{k=1}^{K_1} w_{ik} I_i \; ;$$

$$R_2^C \leq \sum_{i=1}^{M} \sum_{k=K_1+1}^{K_1+K_2} w_{ik} I_i \; .$$

(3.46)

We assume an interior solution for R_1^C and R_2^C that ensures that the inequality constraints in (3.46) become ineffective and can be dropped from consideration. Of the remaining five equality constraints, we substitute four directly into the objective function in (3.45). We append the fifth to the objective function by rewriting the maximization problem in Lagrangean form:

$$\begin{aligned} L = P_q Q &- \sum_{i=1}^{M} r_i I_i - s(L_P + L_1 + L_2) \\ &- P_W^C[\sum_{i=1}^{M} \sum_{k=1}^{K} w_{ik} I_i - R_1^C(L_1) - R_2^C(L_2)] \\ &+ P_R^C R_1^C(L_1) + \lambda[f(I_1,...,I_M, L_P) - Q] \; . \end{aligned}$$

(3.47)

To maximize (3.47) with respect to $Q, I_1,...,I_m, L_p, L_1, L_2$ and λ the firm must satisfy the following first order conditions.

$$P_q - \lambda = 0 \; ;$$

(3.48)

$$- r_i - P_W^C \sum_{k=1}^{K} w_{ik} + \lambda f_i = 0 \; ; \tag{3.49}$$

$$i = 1,...,M \; .$$

$$- s + \lambda f_{L_P} = 0 \; . \tag{3.50}$$

$$- s + P_W^C \frac{dR_1^C}{dL_1} + P_R^C \frac{dR_1^C}{dL_1} = 0 \; . \tag{3.51}$$

$$- s + P_W^C \frac{dR_2^C}{dL_2} = 0 \; . \tag{3.52}$$

$$f(I_1,...,I_M, L_P) - Q = 0 \; . \tag{3.53}$$

Let us rewrite first order condition (3.49) as follows:

$$\lambda f_i = r_i + P_W^C \sum_{k=1}^{K} w_{ik} \; ; \tag{3.54}$$

$$i = 1,...,M.$$

Consider for a moment the implications of (3.54) for the profit maximizing firm. Substituting P_q for λ via equation (3.48), equation (3.54) suggests that the profit maximizing firm chooses each nonlabor input so that the value of the input's marginal product equals its marginal cost. The right hand side of (3.54) reflects the marginal cost of input i. It includes the per unit price of input i as well as the cost of disposing of the waste associated with a unit of the input.

Forming a ratio of equation (3.54) for inputs i and h leads to the following condition that the profit maximizing firm must satisfy:

$$\frac{f_i}{f_h} = \frac{r_i + P_W^C \sum_{k=1}^{K} w_{ik}}{r_h + P_W^C \sum_{k=1}^{K} w_{hk}} \; . \tag{3.55}$$

The result in (3.55) suggests that the firm sets the marginal rate of technical substitution between two inputs equal to the ratio of marginal costs of those inputs.

The implications of equations (3.54) and (3.55) are very similar to those of equations (3.18) and (3.19) discussed under the analysis of the household. The difference is that for the firm the user fee for SWS, P_W^C, affects the marginal cost of each input, whereas for the household the user fee affected the marginal cost of each good. Thus, a major implication of a positive user fee for commercial SWS is that inputs that generate high quantities of waste are, ceteris paribus, less desirable to the firm. This gives the firm incentive to seek out low waste inputs which, in turn, gives manufacturers of the inputs incentive to lower the waste associated with their products. In practice, the cost of disposing of the waste associated with an input might be small relative to the price of the input so that changes in the quantity of waste produced by an input might lead to very small shifts in the demand curve for that input.

Combining first order condition (3.48) with (3.50) suggests that the optimizing firm sets the value of the marginal product of production labor equal to the wage rate or the price of labor. To see this, substitute P_q for λ and then rewrite (3.50) by placing the wage rate, s, on the right hand side of the equality:

$$P_q f_{L_P} = s . \tag{3.56}$$

The two first order conditions represented by equations (3.51) and (3.52) require the firm to choose the number of labor hours spent recycling so that the value of the marginal product of such labor is equal to its marginal cost, the wage rate. To see this, rewrite (3.51) and (3.52) by placing the wage rate on the right hand sides of the equalities:

$$\frac{dR_1^C}{dL_1}(P_W^C + P_R^C) = s ; \tag{3.57}$$

$$\frac{dR_2^C}{dL_2} P_W^C = s . \tag{3.58}$$

The left hand side of equation (3.57) reflects the value of the marginal product of labor spent recycling marketable waste. This value includes the price received by the firm for recycling a unit of waste as well as the user fee for SWS that the firm avoids paying. The left hand side of (3.58) reflects the value of the marginal product of labor spent recycling nonmarketable waste. The value of this marginal product includes only the avoided cost of disposal since the recyclables are not actually sold.

If we divide both sides of equation (3.57) by $(P_W^C + P_R^C)$ and both sides of equation (3.58) by P_W^C, we can easily analyze the effects of a rise in the commercial user fee for SWS.

$$\frac{dR_1^C}{dL_1} = \frac{s}{P_W^C + P_R^C} . \qquad (3.59)$$

$$\frac{dR_2^C}{dL_2} = \frac{s}{P_W^C} . \qquad (3.60)$$

Recall our assumption that as more labor is devoted to recycling, the productivity of the labor declines. Given this assumption, equations (3.59) and (3.60) suggest that more labor will be allocated to recycling tasks in response to an increase in the user fee for SWS, P_W^C. In both equations, as the user fee increases, the marginal product of labor in recycling is equated to a smaller value. This implies that labor used in recycling has risen. Clearly, the higher the user fee the greater are the quantities of waste recycled, at least until the upper bound of recyclable waste is reached.

In conclusion, there are two major consequences of a commercial user fee for SWS. First, waste-intensive inputs become less attractive because their marginal costs rise. Second, the firm finds it worthwhile to increase the amount of labor allocated to recycling tasks.

Assume that the firm's first order conditions for profit maximization, (3.48) through (3.53) can be solved for $I_1,...,I_m$, L_1 and L_2. The solutions, all unconditional demand functions, will be functions of the parameters of the system. Denote these solutions as

$$I_i = H_i(r_1,...,r_M, s, P_R^C, P_W^C, w_{11},...,w_{MK}),$$

$$i = 1,...,M,$$

$$L_1 = H_{L_1}(r_1,...,r_M, s, P_R^C, P_W^C, w_{11},...,w_{MK}),$$

$$L_2 = H_{L_2}(r_1,...,r_M, s, P_R^C, P_W^C, w_{11},...,w_{MK}). \qquad (3.61)$$

Using evident notation from (3.61), the firm's unconditional demand function for SWS, say W_C^d, can now also be expressed as a function of the system parameters:

$$W_C^d = W^C - R_1^C - R_2^C$$

$$= \sum_{i=1}^{M} \sum_{k=1}^{K} w_{ik} H_i(.) - R_1^C[H_{L_1}(.)] - R_2^C[H_{L_2}(.)] \qquad (3.62)$$

$$= F^C(r_1,...,r_M, s, P_R^C, P_W^C, w_{11},...,w_{MK})$$

where $F^c(.)$ is a function of the indicated arguments. Thus the firm's unconditional demand for SWS is a function of nonlabor input prices, the wage rate, the price received for recycling, the commercial user fee for SWS, and the waste associated with each input.

When discussing the empirical model for the firm, we will suggest an estimable form of the firm's demand function for SWS. For the empirical model, we focus on the conditional demand for SWS. The reason is that most firms probably determine their demand for SWS after deciding the quantity of output to produce. In other words, the firm's demand for SWS is probably conditioned on a prespecified level of output.

We can easily show that the firm's conditional demand is a function of the same arguments that appear in $F^c(.)$ plus a prespecified level of output, Q^o; that is,

$$W_{C2}^d = F_2^C(r_1,...,r_M, s, P_R^C, P_W^C, w_{11},...,w_{MK}, Q^o) \qquad (3.63)$$

where $F_2^C(.)$ is the firm's conditional demand function for SWS. To understand why, consider the firm's cost minimization problem, which yields the conditional demand functions. To minimize costs, the firm should maximize equation (3.45) after replacing the choice variable, Q, with a prespecified quantity of output, Q^o. This revised equation (3.45) should be maximized subject to the same constraints as in (3.46), again, after replacing all the unknown Q's in the constraints with known Q^o's.

The firm's cost minimization problem yields the same first order conditions as given by the profit maximization problem with the exception of condition (3.48). The latter need not be satisfied by the firm that minimizes costs. The absence of (3.48) does not affect the substance of the conclusions we draw above. In response to a positive user fee for SWS, the cost minimizing firm has incentive to purchase smaller quantities of inputs that are waste intensive and to increase the quantity of labor assigned to recycling tasks.

NOTES

1. It should become clear that w_i can just as easily be interpreted as the volume of waste.
2. For a citation, see note 1 in Chapter 2.
3. A well-known result is that the Lagrange multiplier associated with an objective function can be interpreted as the change in that objective function when the corresponding constraint is relaxed by a small amount. It follows that our Lagrange multiplier, λ, (see equation (3.12)) can be interpreted as the marginal utility of income.
4. Note that (3.23) and (3.24) also suggest that, everything else equal, the number of hours spent recycling varies inversely with the wage rate, s. This conclusion follows from our assumption that there is no positive utility derived from recycling.
5. See, for example, Langham (1972, pp. 315-22) and Stevens (1988, pp. 285-96).

4. The Residential and Commercial Demand Equations: Some Econometric Issues

4.1 INTRODUCTION

In this chapter we specify the demand equations that we will estimate. These equations are linear versions of the reduced form equations given by (3.32) and (3.63) that are the demand functions for residential and commercial SWS. We specify panel data models of demand. The cross-sectional units of these models represent communities; the interval of time represents months.

The models specified are fixed effects models; that is, the intercepts of the demand functions vary over the communities. The remaining 'slope' coefficients, however, are assumed not to vary.

Our data come from nine communities. These communities are listed in Table 4.1, along with an associated number that we use below for ease of reference. For a complete description of the data set and for summary tables that explain the data obtained from each community, please see Chapter 5.

For communities one, two, four, six and nine, our data represent the sum of residential and commercial demand for SWS. For communities three, five, and eight the data represent only residential demand for SWS. Finally, for community seven, we have data that represent commercial demand for SWS and separate data that represent residential demand for SWS. In this chapter, we describe the procedure we used to efficiently pool all of the available data to estimate both the residential and commercial demand models.

4.2 SPECIFICATIONS OF THE EMPIRICAL MODEL

We first specify the two demand models using general notation. Later, we rewrite the models taking the structure of the available data into account.

Table 4.1 Communities for Which We Have Data

---

1. San Francisco, California
2. The unincorporated parts of Hillsborough County,
   Florida (henceforth Hillsborough County)
3. St. Petersburg, Florida
4. Estherville, Iowa
5. Howard County, Maryland
6. Highbridge, New Jersey
7. Bernalillo County, New Mexico (home of Albuquerque)
8. Seattle, Washington
9. Spokane, Washington

For now consider the general model,

$$y_{it}^R = a_i^R + X_{it}^R B^R + \epsilon_{it}^R$$

$$y_{it}^C = a_i^C + X_{it}^C B^C + \epsilon_{it}^C,$$

$$i = 1,...,n \ ;$$

$$t = 1,...,T_i \ ;$$

(4.1)

where

$$y_{it}^R = \frac{Y_{it}^R}{POP_{it}},$$

$$y_{it}^C = \frac{Y_{it}^C}{E_{it}}.$$

(4.2)

Y_{it}^R and Y_{it}^C are, respectively, residential and commercial demand for SWS in the ith community at time t; POP_{it} is the population of community i at time t; E_{it} is the number of people working in community i at time t; X_{it}^R and X_{it}^C are corresponding vectors of explanatory variables to be presented in detail later; a_i^R and a_i^C are the fixed effects relating to the communities; B^R and B^C are parameter vectors; ε_{it}^R and ε_{it}^C are corresponding disturbance terms.

We assume that ϵ_{it}^{R} is normally and independently distributed over time and over communities, with zero mean and variance σ_R^2. Similarly, we assume that ϵ_{it}^{C} is normally and independently distributed over time and communities, with zero mean and variance σ_C^2.

Let

$$\epsilon_i^R = (\epsilon_{i1}^R,...,\epsilon_{iT_i}^R)'; \quad \epsilon_i^C = (\epsilon_{i1}^C,...,\epsilon_{iT_i}^C)'$$

$$\epsilon^R = (\epsilon_1^{R'},...,\epsilon_n^{R'})'; \quad \epsilon^C = (\epsilon_1^{C'},...,\epsilon_n^{C'})'\ . \tag{4.3}$$

We also assume that ϵ^R and ϵ^C are independent.

In general let e_r be an $r \times 1$ vector of unit elements. Also let

$$y_i^\lambda = (y_{i1}^\lambda,...,y_{iT_i}^\lambda)'; \quad y^\lambda = (y_1^{\lambda'},...,y_n^{\lambda'})'$$

$$X_i^\lambda = (X_{i1}^{\lambda'},...,X_{iT_i}^{\lambda'})'; \quad X^\lambda = (X_1^{\lambda'},...,X_n^{\lambda'})' \tag{4.4}$$

$$Z_i^\lambda = (e_{T_i},X_i^\lambda), \quad \lambda = R, C.$$

Then we assume that the elements of X^C and X^R are nonstochastic, and that

$$\lim T_i^{-1} Z_i^{C'}Z_i^{C} = M_{iC}; \quad \lim T_i^{-1} Z_i^{R'}Z_i^{R} = M_{iR} \tag{4.5}$$

where

$$M_{iC}^{-1},\ M_{iR}^{-1}\ \text{exist}, \tag{4.6}$$

$$i = 1,...,n.$$

4.3 STACKING AND AGGREGATING THE DATA

Using the notation in (4.3) and (4.4), the monthly models in (4.1) can be expressed as

$$y_i^R = (e_{T_i})a_i^R + X_i^R B^R + \epsilon_i^R;$$

$$y_i^C = (e_{T_i})a_i^C + X_i^C B^C + \epsilon_i^C, \qquad (4.7)$$

$$i = 1,...,n.$$

Consider now the limitations imposed by the data. Recall that for communities one, two, four, six and nine, the data relate to the sum of residential and commercial demand for SWS. For these communities, we estimate an aggregated demand equation; that is, one that represents the sum of commercial and residential demands for SWS. A difficulty arises, however, when trying to combine the equations in (4.1). The dependent variable for the equation corresponding to the residential sector is the quantity of residential waste per capita. The corresponding variable for the commercial sector is the quantity of commercial waste per employee. For those communities for which only the sum $Y_{it}^R + Y_{it}^C$ is available, we take the dependent variable to be $(Y_{it}^R + Y_{it}^C)/POP_{it}$. This leads to a scaling problem with respect to the regressors of the model.

To see the issue involved, note that

$$\frac{Y_{it}^R + Y_{it}^C}{POP_{it}} = \frac{Y_{it}^R}{POP_{it}} + \left(\frac{E_{it}}{POP_{it}}\right)\frac{Y_{it}^C}{E_{it}}$$

$$= a_i^R + X_{it}^R B^R + \left(\frac{E_{it}}{POP_{it}}\right)a_i^C + \left(\frac{E_{it}}{POP_{it}}\right)X_{it}^C B^C \qquad (4.8)$$

$$+ \left[\epsilon_{it}^R + \left(\frac{E_{it}}{POP_{it}}\right)\epsilon_{it}^C\right],$$

Clearly, if biases are to be avoided, the employment to population ratio must be applied to the variables of the commercial sector model. The aggregated demand equation can thereby be expressed as

$$\frac{(Y_{it}^R + Y_{it}^C)}{POP_{it}} = a_i^R + X_{it}^R B^R + \left(\frac{E_{it}}{POP_{it}}\right)a_i^C + \left(\frac{E_{it}}{POP_{it}}\right)X_{it}^C B^C \tag{4.9}$$

$$+ \left[\epsilon_{it}^R + \left(\frac{E_{it}}{POP_{it}}\right)\epsilon_{it}^C\right],$$

$t = 1,...,T_i$; $i = 1, 2, 4, 6, 9$.

Let

$$y_{it}^{RC} = \frac{(Y_{it}^R + Y_{it}^C)}{POP_{it}}, \quad y_i^{RC} = (y_{i1}^{RC},...,y_{iT_i}^{RC})', \tag{4.10}$$

$t = 1,...,T_i$; $i = 1, 2, 4, 6, 9$.

Also let

$$d_{it} = \frac{E_{it}}{POP_{it}}, \quad d_i = (d_{i1},...,d_{iT_i})',$$

$$D_i = diag_{t=1}^{T_i}(d_{it}), \tag{4.11}$$

$t = 1,...,T_i$; $i = 1,...,9$.

Then, in the notation of (4.3), (4.4), (4.10) and (4.11), the aggregated model can be expressed as

$$y_i^{RC} = e_{T_i} a_i^R + d_i a_i^C + X_i^R B^R + D_i X_i^C B^C$$

$$+ (\epsilon_i^R + D_i \epsilon_i^C), \tag{4.12}$$

$i = 1, 2, 4, 6, 9$.

The relevant models for the remaining communities are

$$y_i^R = (e_{T_i})a_i^R + X_i^R B^R + \epsilon_i^R, \quad i = 3, 5, 7, 8 ; \tag{4.13}$$

$$y_i^C = (e_{T_i})a_i^C + X_i^C B^C + \epsilon_i^C, \quad i = 7.$$

The models in (4.12) and (4.13) describe our entire data set. We now stack the models and give efficient estimators of the regression parameters. Let

$$y' = (y_1^{RC'}, y_2^{RC'}, y_3^{R'}, y_4^{RC'},$$

$$y_5^{R'}, y_6^{RC'}, y_7^{R'}, y_7^{C'}, y_8^{R'}, y_9^{RC'});$$

$$\gamma' = (a_1^R, a_1^C, a_2^R, a_2^C, a_3^R, a_4^R, a_4^C, \quad (4.14)$$

$$a_5^R, a_6^R, a_6^C, a_7^R, a_7^C, a_8^R, a_9^R, a_9^C, B^{R'}, B^{C'});$$

$$U' = (\epsilon_1^{R'} + (D_{T_1}\epsilon_1^C)', \epsilon_2^{R'} + (D_{T_2}\epsilon_2^C)', \epsilon_3^{R'}, \epsilon_4^{R'} + (D_{T_4}\epsilon_4^C)',$$

$$\epsilon_5^{R'}, \epsilon_6^{R'} + (D_{T_6}\epsilon_6^C)', \epsilon_7^{R'}, \epsilon_7^{C'}, \epsilon_8^{R'}, \epsilon_9^{R'} + (D_{T_9}\epsilon_9^C)')$$

and

$$X = \begin{pmatrix} e_{T_1} & d_1 & 0_{T_1x13} & & X_1^R & D_1X_1^C \\ 0_{T_2x2} & e_{T_2} & d_2 & 0_{T_2x11} & X_2^R & D_2X_2^C \\ 0_{T_3x4} & e_{T_3} & 0_{T_3x10} & & X_3^R & 0_{T_3xKc} \\ 0_{T_4x5} & e_{T_4} & d_4 & 0_{T_4x8} & X_4^R & D_4X_4^C \\ 0_{T_5x7} & e_{T_5} & 0_{T_5x7} & & X_5^R & 0_{T_5xKc} \\ 0_{T_6x8} & e_{T_6} & d_6 & 0_{T_6x5} & X_6^R & D_6X_6^C \\ 0_{T_7x10} & e_{T_7} & 0_{T_7x4} & & X_7^R & 0_{T_7xKc} \\ 0_{T_7x11} & e_{T_7} & 0_{T_7x3} & & 0_{T_7xKr} & X_7^C \\ 0_{T_8x12} & e_{T_8} & 0_{T_8x2} & & X_8^R & 0_{T_8xKc} \\ 0_{T_9x13} & e_{T_9} & d_9 & & X_9^R & D_9X_9^C \end{pmatrix} \quad (4.15)$$

where 0_{Tixr} is a T_i by r matrix of zeroes and Kr and Kc are the number of

rows of B^R and B^C respectively.
Let

$$P^l = [\sigma_R^2 + \left(\frac{E_{11}}{POP_{11}}\right)^2 \sigma_C^2, \ \sigma_R^2 + \left(\frac{E_{12}}{POP_{12}}\right)^2 \sigma_C^2, \ ...,$$

$$\sigma_R^2 + \left(\frac{E_{1T_1}}{POP_{1T_1}}\right)^2 \sigma_C^2,$$

$$\sigma_R^2 + \left(\frac{E_{21}}{POP_{21}}\right)^2 \sigma_C^2, \ \sigma_R^2 + \left(\frac{E_{22}}{POP_{22}}\right)^2 \sigma_C^2, \ ...,$$

$$\sigma_R^2 + \left(\frac{E_{2T_2}}{POP_{2T_2}}\right)^2 \sigma_C^2,$$

$$\sigma_R^2 \, e_{T_3}^l \, ,$$

$$\sigma_R^2 + \left(\frac{E_{41}}{POP_{41}}\right)^2 \sigma_C^2, \ \sigma_R^2 + \left(\frac{E_{42}}{POP_{42}}\right)^2 \sigma_C^2, \ ...,$$

$$\sigma_R^2 + \left(\frac{E_{4T_4}}{POP_{4T_4}}\right)^2 \sigma_C^2,$$

$$\sigma_R^2 \, e_{T_5}^l \, ,$$

$$\sigma_R^2 + \left(\frac{E_{61}}{POP_{61}}\right)^2 \sigma_C^2, \ \sigma_R^2 + \left(\frac{E_{62}}{POP_{62}}\right)^2 \sigma_C^2, \ ...,$$

$$\sigma_R^2 + \left(\frac{E_{6T_6}}{POP_{6T_6}}\right)^2 \sigma_C^2,$$

$$\sigma_R^2 \, e_{T_7}^l \, ,$$

$$\sigma_C^2 \, e_{T_7}^l \, ,$$

$$\sigma_R^2 \, e_{T_8}^l \, ,$$

(4.16)

$$\sigma_R^2 + \left(\frac{E_{91}}{POP_{91}}\right)^2 \sigma_C^2, \; \sigma_R^2 + \left(\frac{E_{92}}{POP_{92}}\right)^2 \sigma_C^2, \; ...,$$

$$\sigma_R^2 + \left(\frac{E_{9T_9}}{POP_{9T_9}}\right)^2 \sigma_C^2] \;\; {}_{1 \; X \; T_1 \; + \; ... \; + \; T_9} \; .$$

Let $T = T_1 + ... + T_9$. Then $V_U = \mathrm{diag}_{i=1}^{T} (P_i)$ where P_i is the ith element of P. Given the notation in (4.14), (4.15) and (4.16), the models in (4.12) and (4.13) can be expressed as

$$y = X \gamma + U, \tag{4.17}$$

where

$$E(U) = 0, \quad E(UU') = V_U \; .$$

4.4 METHOD OF ESTIMATION

Since the model's error terms are heteroskedastic, a consistent and best linear unbiased estimator of our coefficient vector, γ, is the generalized least squares (GLS) estimator, γ^G , where

$$\gamma^G = (X' V_U^{-1} X)^{-1} X' V_U^{-1} y \; . \tag{4.18}$$

This GLS estimator is also efficient under certain conditions.

The estimator is a function of V_u which is, in turn, a function of σ_R^2, σ_C^2, POP_{it} and E_{it}; $i = 1,...,9$; $t = 1,...,T_i$. Both σ_R^2 and σ_C^2 are unknown parameters. Therefore a feasible GLS estimator was used. The feasible GLS estimator was based on estimates of σ_R^2 and σ_C^2 which were calculated in the following way. Recall that the data for communities three, five, seven and eight represent residential demand only. The variance of the error terms corresponding to these demand equations is assumed to be the same across all four communities. Thus, we stacked the demand equations for these communities and estimated them by ordinary least squares (OLS). The estimate of σ_R^2 was then calculated as the error sum of squares divided by $(T_R - K_R)$ where $T_R = (T_3+T_5+T_7+T_8)$ and K_R is the number of parameters in the residential model, including the number of fixed effects. It turned out that $\hat{\sigma}_R^2 = 0.09233$; $T_R = 324$ and $K_R = 12$. The estimate of σ_C^2 was calculated following a similar procedure. We have data that represent commercial demand for SWS for one community only: community seven. The commercial demand equation for this community was estimated by OLS. Then $\hat{\sigma}_C^2$ was calculated as the error sum of squares divided by $(T_7 - K_C)$ where K_C is the number of parameters in the commercial model, including the intercept term for community seven. Our results were that $\hat{\sigma}_C^2 = 0.35268$ given that $T_7 = 81$ and $K_C = 6$.

The tests of hypotheses described in Chapters 6 and 7 are based on the assumption that our data are such that the GLS estimator and the feasible GLS estimator have the same asymptotic normal distribution. Conditions ensuring that they converge in probability and therefore in distribution are given in Schmidt (1976, p. 71).

4.5 SUMMARY

In this chapter we have specified empirical models of residential and commercial demand for SWS. The models consist of three different equations: one that represents residential demand, another that represents commercial demand and a third that represents the sum of residential and commercial demand. Stacking and aggregating the data suggest that the model has a heteroskedastic error structure. Thus we outline a feasible GLS estimator of the coefficient vector.

5. A Description of the Data and Details of the Empirical Model

5.1 AN OVERVIEW

There is no national trade association or federal government agency that keeps track of the quantities of residential and commercial wastes collected in the US. Thus, to obtain quantity and price data relating to SWS, we found it necessary to contact community waste agencies directly. We made efforts to contact waste agencies in all the US communities where we knew residents were paying volume-based user fees for SWS. We succeeded in obtaining data for five such communities. We also collected data from waste agencies in four communities whose residents do not pay volume-based user fees. The latter communities were selected from a list of solid waste agencies with whom we had already made contact under the mistaken belief that they were charging user fees. Our familiarity with their data allowed us to easily select four successful candidates for inclusion in the final data set.

Precisely because our data came from many different sources, there are inconsistencies in the types of waste that it represents. For example, military bases in some communities have their own landfills. In such communities, the base's waste quantities would not be represented by our data which measures the waste sent to municipal landfills. Exacerbating the data inconsistencies are the legal and institutional factors that affect the quantities that residents and businesses discard. Such factors vary from community to community and include, for example, the legality of burning refuse in residential areas or the ease with which residents can recycle their waste. Such variation among communities was unavoidable, although we have made efforts to choose communities with limited differences. In this chapter, we will highlight the most important inconsistencies in the data related to waste quantities.

We have also collected data that relate to community characteristics, such as population and per capita income, and other data that relate to regional prices such as the market price of old newspapers. These data have come from a wide variety of sources. We will describe these data, their potential shortcomings and their origins.

In the final section of this chapter we give a detailed version of the empirical model outlined in Chapter 4. The detail explains how we have tailored the model to fit the data.

5.2 DATA RELATED TO WASTE QUANTITIES AND PRICES OF SWS

Five communities that currently charge households volume-based user fees also keep records on the weight of waste collected and the prices charged for SWS. The household user fee variants in these communities are as follows. Estherville, Iowa requires refuse to be set out for pick up in specially marked, positively priced plastic bags. Highbridge, New Jersey requires a positively priced sticker to be affixed to refuse containers. The three remaining 'user fee communities,' San Francisco, Seattle, and Spokane, charge households for SWS according to the number of pre-specified refuse containers set out.

In addition to data for user fee communities, we also want data for communities that do not charge a user fee. Such communities either charge a flat fee for SWS or deduct the cost of SWS from property taxes. Data for such communities represent the demand for SWS when the marginal price of SWS is zero. One such community that keeps separate track of the weight of both household and commercial waste collected is Bernalillo County, New Mexico. Three additional communities that do not charge households user fees but do keep track of the weight of waste are St. Petersburg, Florida; Hillsborough County, Florida; and Howard County, Maryland.

We have constructed a data set consisting of information from all of these locales. Table 5.1 indicates the dates and type of data which are available for each of the communities. All data are monthly.

All nine communities measured the quantity of waste collected in tons. In all cases, the tonnage data reflect the total quantity of residential, commercial or mixed waste collected in a community. In other words, we only included data in the sample when its providers could assure us that the data reflected the quantity of waste collected from all the households in a community and/or from all the relevant commercial establishments.[1] This led to the exclusion of data provided by only one of several collection agencies operating in a community. This exclusion, as well as a lack of cooperation from the characteristically competitive private collection companies, left us with data for nine communities whose waste collection agencies were municipally operated or controlled.

Table 5.1 Data Collected				
Community	Volume-Based User Fee Charged for Residential SWS	Residential Waste Data Collected	Commercial Waste Data Collected	Data Collected on the Sum of Residential and Commercial Waste
San Francisco	Yes			1/80- 9/88
Hills-borough County	No			1/84- 6/84 8/84- 2/88
St. Petersburg	No	1/80-12/88		
Estherville	Yes			7/84-11/88
Howard County	No	7/87- 7/89		
Highbridge	Yes			1/88-12/88
Bernalillo County	No	7/82- 5/89	9/82- 5/89	
Seattle	Yes	1/80-12/88		
Spokane	Yes			1/86-12/88

5.21 The Residential Price of SWS

As mentioned, residents in five of the nine communities paid volume-based user fees for SWS. Typically, the actual fee that the resident paid was a matter over which there was some choice. For example, a San Francisco resident could pay extra to receive two collection visits each week rather than one, or to receive backyard rather than curbside collection. We followed a simple rule in constructing the price variable for residential SWS. We always considered the price of the least expensive SWS as the price variable. We reasoned that it was the 'no-frills' service for which households had the least elastic demand. Thus, a measure of the household's response to this no-frills price should be a lower bound of the elasticity of demand for SWS.

The price variable for residential SWS was defined as the price per 30- to 32-gallon container for once-a-week curbside collection service. This rate was calculated as the price paid by the household for the weekly removal of three 30- to 32- gallon containers of refuse, divided by three. To get an idea of how the user fees varied over time within the communities see Table A.1 in Appendix A.

In one user fee community, Estherville, residents were charged a flat fee in addition to the volume-based user fee. In that case, we simply ignored the flat fee since, except for minor income effects, it should not have affected the consumer's decision regarding the quantity of waste to discard. For the four communities whose waste agencies did not charge residents a volume-based user fee, the residential price of SWS was taken to be zero.

5.22 The Commercial Price of SWS

All the communities for which we have data relating to commercial waste quantities charged commercial customers volume-based user fees. As mentioned in Chapter 1, unlike residential user fees, commercial user fees are the norm in the US.

The type of container in which a business may place its refuse varies, and frequently the price of SWS varies according to the type as well as the size of container.[2] For example a roll-off refuse container is designed so that rather than being emptied into a refuse truck, the entire container is rolled onto the back of a truck. The price of SWS is typically higher for a roll-off container than for an ordinary refuse container, which we shall refer to as a dumpster. The quality of a dumpster can also vary. As a result, the price of SWS sometimes depends on the 'class' of dumpster.

Two of the smaller communities represented by our data set offered only dumpster service and a single quality of dumpster at that. Waste officials associated with bigger city waste agencies suggested that the dumpster service is the most popular. Conversations with waste officials also suggested that most of the businesses that subscribe to dumpster service receive two collection visits each week. These practices suggested that the commercial price of SWS should be taken as the price of having the least expensive available dumpster emptied twice a week. Specifically, the commercial price was taken to be the weekly rate per cubic yard of dumpster capacity for two pick-ups per week. This price was calculated as the price of the largest (but lowest quality) dumpster available divided by the number of cubic yards in that dumpster. To get an idea of how the commercial user fees varied over time within the communities see Table A.1 in Appendix A.

5.3 THE FUNDING OF SWS IN NON-USER FEE COMMUNITIES

Waste agencies in the four communities that were not subject to volume-based user fees financed their SWS by other means. Hillsborough and Bernalillo County residents paid monthly or annual flat fees. In Howard County, funds came directly from local property taxes. The most complicated case is St. Petersburg. Residents there paid a flat fee for SWS until October of 1981. At that time, the St. Petersburg waste collection agency began a pilot program. A sample of residents began receiving mechanized collection service; garbage trucks equipped with mechanical arms picked up standardized refuse containers. The waste agency provided the sample of households with standardized 90-gallon refuse containers. To receive any more than one 90-gallon container these residents were required to pay extra. Due to the large size of the container and the fact that all households were visited twice each week for refuse collection, we did not consider the fee per 90-gallon container a volume-based user fee. The minimum service level probably did not constrain many households. In addition to this, much of the data for St. Petersburg reflect the quantities of waste set out by households who did not participate in the pilot program; although the number of participants did grow from the end of 1981 through the period for which we have data. Thus, we considered St. Petersburg a flat fee community.

5.4 INCONSISTENCIES WITHIN THE QUANTITY DATA

Several inconsistencies exist within the waste tonnage data. One source of inconsistency is the way the various waste agencies defined residential and commercial waste. At issue is the category to which apartment waste belongs. Some communities defined apartment waste as commercial while others defined it as residential. Inconsistencies also arise from different laws among communities that determined the types of waste a waste agency collects. For example, one community permitted the burning of refuse; another required special handling of bulky items such as sofas or refrigerators; others had bottle bills; and so on. Finally, inconsistent quantity data result from variations in the no-frills level of service - that is, the lowest priced SWS - and in the minimum number of containers for whose removal residents are required to pay. A perfect data set would include only communities that provided one particular no-frills service level and a uniform minimum subscription level.

5.41 Apartment Waste

Recall that for four communities, we have quantity data that reflect only the waste discarded by households or 'residents.' In two of these communities, Seattle and Howard County, the waste collection agents considered the waste generated by apartment dwellers as residential waste. Unfortunately, in the other two, Bernalillo County and St. Petersburg, any apartment waste that was placed in dumpsters was *not* considered residential waste but, instead, commercial waste.

This brings up a different point. Apartment dwellers in the user fee communities were not always subject to user fees for SWS. Residents of apartment buildings whose owners paid for dumpster service did not have to pay a fee for refuse collection. It seems reasonable to assume that the likelihood of dumpster service increases with the number of dwelling units per apartment. Assuming that the larger the city, the higher density its housing units, we should be concerned that the two larger user fee cities, San Francisco and Seattle, might have a high proportion of residents not subject to user fees. This concern is valid for San Francisco. Approximately one-third of the population there lived in apartment buildings of ten or more dwelling units in 1988 (*1988 Revisions* ... 1988, pp. 1-3, 3-75). The concern is not valid for Seattle, however, since apartment residents there were charged volume-based user fees, although the fees were tied to the number of housing units as well (Skumatz 1990, p.3). Fortunately, the inclusion of waste quantities generated by residents not subject to user fees in our data should reduce the response of demand to the price of SWS rather than exaggerate it.

5.42 Laws and Policies That Affect Residential Waste Quantities

Many different laws and policies affect the type and quantity of waste that an agency collects. Table 5.2 presents the laws we expected to affect the quantity of residential waste collected.

Table 5.2 Laws and Policies That Affect Residential Waste Quantities (Part 1)			
Community	Open Burning of Refuse Permitted	Acceptable Refuse Container Size, in gallons	Must Change Subscription Level to Vary Number of Containers
San Francisco	No	32	Yes
Hillsborough County	No	10 - 30	N/A[a]
St. Petersburg	No	10 - 30 until 10/81, then 90 for some residents	N/A[a]
Estherville	Yes	≈ 30[b]	No
Howard County	No	20	N/A[a]
Highbridge	No	30	No
Bernalillo County	No	≈ 30[b]	N/A[a]
Seattle	No	32	Yes
Spokane	No	10 - 32	Yes

a. The waste collection agency does not charge residents a volume-based user fee in this community.
b. Residents place refuse in plastic bags.

Community	Yard Waste Treated as Any Other Waste	Community Subject to a Bottle Bill[c]	Bulky Items Treated as Any Other Waste	Newspapers Treated as Any Other Waste
Table 5.2 Laws and Policies That Affect Residential Waste Quantities (Part 2)				
San Francisco	Yes	Yes, as of 10/87	No	Yes
Hills-borough County	Yes	No	No	Yes
St. Petersburg	only if pre-sented in small quan-tities	No	Yes	Yes
Estherville	Yes[d]	Yes	No	Yes
Howard County	Yes	No	No	Yes
Highbridge	No	No	Yes	No
Bernalillo County	Yes	No	No	Yes
Seattle	Yes	No	No	Yes
Spokane	No	No	Yes	Yes

c. In other words, residents receive some sort of payment for turning in certain beverage bottles.
d. The first 12 months of quantity data for Estherville does not reflect the weight of residential yard waste.

5.43 Laws and Policies That Affect Commercial Waste Quantities

There are also policies that affect the types of commercial wastes collected. A review of the most important of these is presented in Table 5.3.

Community	Construction and Demolition Debris	Industrial Process Waste	Federal Government Office Waste
Table 5.3 Special Types of Commercial Waste Excluded from the Waste Tonnage Data			
San Francisco	Excluded	Some, not all, included	Excluded
Hillsborough County	Some, not all, included	Some, not all, included	Included
Estherville	Excluded	Some, not all, included	Unknown
Highbridge	Excluded	Excluded	Excluded
Bernalillo County	A small portion included	Some, not all, included	Some, not all, included
Spokane	A small portion included	Excluded	Very little included, if any

5.44 The No-Frills SWS Level

The quantity data that represent residential waste might vary over communities due to variation in the lowest priced or no-frills SWS level. Two service characteristics that might affect waste quantities are the frequency of collection visits and the location of collection - backyard versus curbside. Research reviewed in Chapter 2 by Kemper and Quigley certainly suggests this. These service levels might affect waste quantities by affecting the ease with which waste can be discarded and by affecting the build-up of waste quantities on residential premises.

Waste collection agents in eight of the nine communities provided curbside or alley collection to residents. This was the only service level offered in five of the communities. It was the lowest priced service level in three others, San Francisco, Estherville and St. Petersburg. All residents received backyard collection in Seattle. Residents in four of the five user fee communities and in Bernalillo County had their waste removed once a week. Only in San Francisco did residents have a choice over the frequency of service. San Franciscans could pay twice the user fee for two collections each week. Households in the three remaining communities, Hillsborough County, Howard County and St. Petersburg, all of which are non-user fee communities, had their waste removed twice each week.

5.45 Minimum and Maximum Service Levels

The five user fee communities all had different requirements regarding the minimum number of refuse containers households were required to pay for. None allowed households to escape payment altogether. Seattle and Estherville came the closest to such a policy. Residents in Seattle could subscribe to 'zero can' service - that is, no service - for a very low monthly charge. Residents in Estherville were not required to purchase any plastic bags, the equivalent of a zero subscription level. However, as mentioned, Estherville waste authorities charged all households a monthly flat fee to cover fixed costs.

Two other user fee communities, San Francisco and Highbridge, required residents to pay for the removal of at least one container of refuse each week. In Highbridge, where a sticker must be affixed to each 32-gallon refuse container, this was accomplished by requiring residents to buy 52 stickers each year. San Francisco simply had a minimum subscription level of one container per week. Exceptions were made in San Francisco but only for senior citizens who could demonstrate that they discarded less than 20 gallons per week.

The final user fee city, Spokane, required households to pay for the removal of two containers each week. Again, there were exceptions to the rule. In this case, to receive the single can rate a household was required to get approval from the waste agency. Presumably, approval was only given to households who could demonstrate that they discarded less than 32 gallons each week.

The flat fee communities did not have minimum subscription levels.[3] In fact in Bernalillo County, residents could haul their own waste rather than participate in the city collection service.[4] Howard County had a *maximum* number of four containers that would be collected each visit - recall that visits were made twice each week. Residents in St. Petersburg who did not receive mechanical service were subject to a maximum number of two containers that would be collected each visit - again, recall that visits were made twice per week.

5.46 The Ease of Recycling

A final institutional factor that might affect the quantity of residential waste is the ease with which households can recycle their waste. Only one of the communities, Estherville, had no recycling programs. Most of the communities had drop-off centers - places that accept recyclable items but to which residents must transport their own recyclables. Three communities, Seattle, Highbridge and Howard County, began curbside recycling programs in the final months of the periods for which we have data. Curbside recycling programs usually involve the collection of recyclables from the householder's curbside or alley. During most of the time and in most of the communities for which we have data, curbside recycling programs were nonexistent or were only in the planning stages. Interestingly, Bernalillo County and Spokane planned to begin curbside recycling programs in the months following the periods for which we have data.

5.47 Conclusion

Most of the inconsistencies reflected by the quantity data occur over communities, not time. Recall that our model is a fixed-effects one in which the intercepts vary over communities. In other words, for each community the model allows the height of the demand curve for SWS to differ. Our assumption is that the inconsistencies will be reflected by the different heights or, more technically, the community intercepts. We assume that the inconsistencies will not affect the slope of the demand curve for SWS. The slope coefficient estimates are the ones whose interpretations are of interest. In Chapter 7, we will present the results of a statistical test of the latter assumption.

5.5 DATA RELATED TO COMMUNITY CHARACTERISTICS AND REGIONAL PRICES

Table 5.4 summarizes the data related to community characteristics and regional prices. These data are usually monthly, bi-monthly or yearly. Some of the data, such as that related to population, correspond to the individual communities. Most of the data, however, correspond to broader regions such as the southern or the northeastern US. These regions are described in detail in Table 5.4. To get an idea of the actual values of the variables described in Table 5.4, see Table A.2 in Appendix A.

Variable	Description	Source[a]	Fre-quency	Geo-gra-phic Units[b]
Consum-er Price Index (CPI)	Regional CPI for all items, for all urban consumers (Base 1982 - 1984)	Bureau of Labor Statistics (BLS)	Bi-monthly or semi-yearly[c]	Type 1 Re-gions[d]
Producer Price Index (PPI)	National PPI for all commodities (Base 1982)	BLS	Monthly	US
Age Dis-tribution of the Popu-lation	Percentage of the population aged 18 to 49	Survey of Buying Power (SBP)	Yearly	Type 1 Com-muni-ties[e]
House-hold Income	Median household disposable income - personal income less personal taxes	SBP	Yearly	Type 1 Com-muni-ties
House-hold Size	Total population divided by the number of households	SBP	Yearly	Type 2 Com-muni-ties[f]
Price of Old News-papers	The midpoint of price ranges of the price per short ton for newspaper - reflects the paper mill's purchase price less delivery charges	Official Board Markets - The Yellow Sheet	Monthly	Type 2 Re-gions[g]

Table 5.4 Data Related to Community Characteristics and Regional Prices

Price of Old Corrugated Containers	The midpoint of price ranges of the price per short ton for corrugated containers - reflects the paper mill's purchase price less delivery charges	Official Board Markets - The Yellow Sheet	Monthly	Type 3 Regions[h]
Population	Population	Local Planning Commissions or Depts of Labor	Yearly	Communities of the sample
Employment	Total nonfederal, nonagricultural, nonmining, non-construction and nonmanufacturing employment	BLS - ES-202	Yearly	Communities of the sample[i]
Square Miles	The total number of square miles of dry land and land temporarily or partially covered by water	Census Bureau	Not Applicable	Communities of the sample
Precipitation	The total inches of precipitation	National Climatic Data Center	Monthly[j]	Type 3 Communities[k]
Average Temperature	The mean temperature in degrees Fahrenheit	National Climatic Data Center	Monthly	Type 3 Communities
Diesel Fuel Prices	The average wholesale price per gallon of no. 2 distillate (diesel fuel)	Dept of Energy	Monthly	Type 4 Regions[l]

Wages	Annual wages per employee in standard industrial classification 4953 - refuse systems	BLS - ES-202	Yearly	States

a. See Appendix B for complete citations.
b. This column gives the geographic units to which the data apply.
c. For the period prior to 1987, this data is bimonthly. Beginning January 1987, this data is semiannual for Seattle and monthly for all the other regions.
d. *Type 1 Regions* - San Francisco-Oakland-San Jose, CA; North central urban (Estherville); South urban (Hillsborough County, St. Petersburg, Howard County); Northeast urban (Highbridge); West urban (Bernalillo County, Spokane); Seattle-Tacoma, WA.
   The names in parentheses are the communities for which the region's data were used. No name is given when the connection is obvious. This comment applies to the following regions and communities as well.
e. *Type 1 Communities* - San Francisco; Hillsborough County; St. Petersburg; Emmet County, Iowa (Estherville); Howard County; Hunterdon County (Highbridge); Bernalillo County; Seattle; Spokane.
f. *Type 2 Communities* - Same as Type 1 Communities except instead of Hillsborough County, data is for Hillsborough County minus Tampa.
g. *Type 2 Regions* - New York Area (Highbridge); Southeast (Hillsborough, St. Petersburg, Howard County); Chicago (Estherville); San Francisco/Los Angeles area (San Francisco, Bernalillo County, Seattle, Spokane).
h. *Type 3 Regions* - Same as Type 2 Regions less Chicago. The price of old corrugated containers for Estherville was unavailable so we substituted the price of old newspapers.
i. Data for Spokane, Hillsborough County (the unincorporated parts), Estherville and Highbridge were estimated from county numbers by multiplying the percentage of county employment occurring in each community in 1980 by the total county employment in later years.
j. Three data points were missing. They were estimated using data from earlier and later years.
k. *Type 3 Communities* - San Francisco; Tampa, Florida (Hillsborough, St. Petersburg); Estherville; Clarksville, Maryland (Howard County); Lambertville, New Jersey (Highbridge); Albuquerque (Bernalillo County); Seattle; Spokane.
l. *Type 4 Regions* - Middle southeast (Howard County, Highbridge); South (Hillsborough County, St. Petersburg); Midwest (Estherville); Southwest

(Bernalillo County); West and others (San Francisco, Seattle, Spokane).

Note that the regional CPIs do not reflect the level of prices in one community relative to another.

> The indexes cannot be used to determine relative living costs. An individual geographic area index measures how much prices have changed in that particular area over a specific time period. It does not show whether prices or living costs are higher or lower in that area relative to another (US Department of Labor, Bulletin 2285, 1988, p. 158).

Fortunately, because our model is a fixed effects one, these differences in prices over communities should affect only the constant terms.

The PPI is a national one. Thus, all commercial prices are deflated by this national index rather than by more specific regional indexes.

The geographic units for which the age distribution of the population, household income and household size were available are less specific than we desired. These data were not available for the two small communities, Estherville and Highbridge, so we used county level data for them instead. Similarly, data relating to Hillsborough County were for the entire county instead of just the unincorporated parts.

The geographic regions for which the prices of old newspapers and corrugated containers were available are very limited. Fortunately, the available regions matched up nicely with five of our communities. However, we were forced to use San Francisco-Los Angeles area prices for Bernalillo County, Seattle and Spokane. Perhaps even less appropriate was our use of Chicago prices for Estherville. Furthermore, since the prices of old corrugated containers in Chicago were unavailable, we used Chicago prices of old newspapers as a proxy. We did not use prices of old corrugated containers for some other region as the proxy because we found that within each region, the movement of the two price series paralleled one another more closely than the movement of prices of old corrugated containers across regions.

Next to the waste quantity data, the population data are probably the least uniform across communities. These data come from nine different sources, one for each community. Thus there is a high probability that the methods by which the figures are calculated differ among communities. This diversity of sources was unavoidable given our need for data that reflected only the population of the specific community and whose frequency was at least yearly.

The employment data are derived from the ES-202 data series. This series consists of data collected by the BLS to record federal unemployment insurance payments. Thus the employment data are based on universes rather than samples. We excluded federal, agricultural, mining, construction and manufacturing employment since the waste quantity data usually did not reflect the waste discarded by these employers (see Table 5.3). The employment data thus reflect the number of workers in transportation, public utilities, the wholesale and retail trades, finance, insurance, real estate, services, public administration and any 'non-classifiable' trades. The waste from these trades is almost always reflected by the tonnage data.

We chose to measure the price that waste collection agencies paid for diesel fuel for garbage trucks as the wholesale rather than the retail price. Typically, with truck fleets in excess of five or so vehicles, the collection agency will purchase diesel fuel at wholesale prices. We suspect that the truck fleets operating within all nine communities were larger than five vehicles.

Finally, we chose the average wages earned in the 'refuse systems' industry to reflect wages paid by waste collection and disposal agencies. We considered using the wages earned by workers in 'trucking, local and long distance,' because many of the workers associated with SWS are short distance or local truckers. However, the aggregation of the wages earned by long distance truckers with those earned by local truckers reduced the desirability of this data. As a result, we adopted the wage data related to the 'refuse systems' industry.

5.6 DETAILS OF THE EMPIRICAL MODEL

In Chapter 4, we outlined three empirical demand models that can be estimated with our data (see equations (4.12) and (4.13)). In the preceding sections of this chapter we discussed the details of the data collected. We now present versions of the demand models that reflect the level of detail suggested by the preceding discussion.

Consider the following versions of the models represented by (4.12) and (4.13),

$$\frac{Y_{it}^R}{POP_{it}} = a_i^R + \frac{X_{it1}^R}{CPI_{it}}b_1^R + \frac{X_{it2}^R}{CPI_{it}}b_2^R + X_{it3}^R b_3^R$$

$$+ \frac{X_{it4}^R}{CPI_{it}}b_4^R + X_{it5}^R b_5^R + X_{it6}^R b_6^R + X_{it7}^R b_7^R$$

$$+ X_{it8}^R b_8^R + \epsilon_{it}^R \ ,$$

(5.1)

$t = 1,...,T_i, \ \ i = 3,5,7,8;$

$$\frac{Y_{it}^C}{E_{it}} = a_i^C + \frac{X_{it1}^C}{PPI_t}b_1^C + X_{it2}^C b_2^C + \frac{X_{it3}^C}{PPI_t}b_3^C + X_{it4}^C b_4^C$$

$$+ X_{it5}^C b_5^C + \epsilon_{it}^C \ ,$$

(5.2)

$t = 1,...,T_i, \ \ i = 7;$

$$\frac{(Y_{it}^R + Y_{it}^C)}{POP_{it}} = a_i^R + \left(\frac{E_{it}}{POP_{it}}\right)a_i^C + \frac{X_{it1}^R}{CPI_{it}}b_1^R + \frac{X_{it2}^R}{CPI_{it}}b_2^R$$

$$+ X_{it3}^R b_3^R + \frac{X_{it4}^R}{CPI_{it}}b_4^R + X_{it5}^R b_5^R + X_{it6}^R b_6^R$$

$$+ X_{it7}^R b_7^R + X_{it8}^R b_8^R$$

$$+ \left(\frac{E_{it}}{POP_{it}}\right)\frac{X_{it1}^C}{PPI_t}b_1^C + \left(\frac{E_{it}}{POP_{it}}\right)X_{it2}^C b_2^C$$

$$+ \left(\frac{E_{it}}{POP_{it}}\right)\frac{X_{it3}^C}{PPI_t}b_3^C + \left(\frac{E_{it}}{POP_{it}}\right)X_{it4}^C b_4^C$$

$$+ \left(\frac{E_{it}}{POP_{it}}\right)X_{it5}^C b_5^C$$

$$+ \epsilon_{it}^R + \left(\frac{E_{it}}{POP_{it}}\right)\epsilon_{it}^C \ ,$$

(5.3)

$t = 1,...,T_i, \ \ i = 1,2,4,6,9;$

where

Y_{it}^R	= the number of pounds of residential waste disposed of per day in community i in month t;
Y_{it}^C	= the number of pounds of commercial waste disposed of per day in community i in month t;
POP_{it}	= the number of people living in community i in month t;
a_i^R	= the fixed effect for community i related to the residential demand equation;
a_i^C	= the fixed effect for community i related to the commercial demand equation;
E_{it}	= the number of people for whom community i is their place of employment in month t;
X_{it1}^R	= the residential volume-based user fee per 32-gallon container in community i in month t;
X_{it2}^R	= disposable income per household in community i in month t;
X_{it3}^R	= the population per square mile in community i in month t;
X_{it4}^R	= the six-month average of the market price paid by paper mills for old newspapers in community i during the six months prior to month t;
X_{it5}^R	= the percentage of the population aged 18 to 49 in community i in month t;
X_{it6}^R	= the mean temperature in degrees Fahrenheit in community i in month t;
X_{it7}^R	= the number of inches of precipitation in community i in month t;
X_{it8}^R	= the average number of persons per household in community i in month t;
CPI_{it}	= the regional consumer price index applicable to community i in month t;
X_{it1}^C	= the weekly commercial volume-based user fee per cubic yard of dumpster capacity in community i in month t;
X_{it2}^C	= the population per square mile in community i in month t;
X_{it3}^C	= the six-month average of the market price paid by paper mills for old corrugated containers in community i during the six months prior to month t;
X_{it4}^C	= the mean temperature in degrees Fahrenheit in community i in month t;
X_{it5}^C	= the number of inches of precipitation in community i in month t;
PPI_t	= the national producer price index in month t;
$b_1^R, b_2^R, ..., b_8^R$	= the coefficients that correspond to the residential regressors;
$b_1^C, b_2^C, ..., b_5^C$	= the coefficients that correspond to the commercial regressors;

ε_{it}^{R} = the disturbance term corresponding to the residential demand model;

ε_{it}^{C} = the disturbance term corresponding to the commercial demand model.

Equations (5.1) through (5.3) represent the demand equations that we have estimated.

NOTES

1. By relevant commercial establishments we mean all those besides heavy industry, construction and demolition businesses and federal government offices. A discussion of the exclusion of data related to the waste quantities discarded by these businesses is given below. This statement is not quite true for the Highbridge data, which reflect the sum of residential and commercial waste quantities. The statement applies to the residential part of the data but not to the commercial. That is, the Highbridge data do not reflect all of the commercial waste discarded in that locale. (They do reflect all of the residential waste discarded.) The commercial waste quantities not reflected are those collected by non-municipal haulers that were operating in Highbridge during the time period for which we have data. Because Highbridge is a suburban community that is almost exclusively residential, we regarded the less-than-perfect data as acceptable.
2. In the largest of the communities in our data set, San Francisco, commercial establishments could pay for SWS according to a rate schedule based on items as specific as the average labor and time required to empty a refuse container. A rate was determined for each individual establishment as follows. The commercial establishment chose the number of waste truck crew members they wished to pay for. Then the collection agency would actually count the minutes that that crew required, on average, to collect the refuse. The average weight of the waste was also calculated based on an initial two or three visits. The commercial user fee charged could then be very precisely calculated by inserting the number of crew members, the minutes required for collection and the average weight of the contents of the refuse container into a formula. These individualized user fees for commercial SWS were difficult to accurately represent by a variable. Fortunately for our analysis, businesses in San Francisco could also pay for SWS according to a predetermined rate schedule. This schedule was based on average crew sizes and average minutes required to collect refuse. Thus, when defining our price variable we relied on this predetermined rate schedule.
3. There is one exception - the residents in the pilot program for mechanized refuse pick-up in St. Petersburg were subject to a minimum subscription level. As previously discussed, these residents were required to rent at least one 90-gallon container, which was then emptied twice per week.
4. The self-hauled waste quantities are reflected by Bernalillo's quantity data.

6. How Waste Quantities Respond to a User Fee for SWS

6.1 AN OVERVIEW

We have estimated the residential and commercial demand models for SWS with the complete data set and by a generalized least squares (GLS) method of estimation. In this chapter we will interpret the results of the estimation. Specifically, we will investigate the response of residential waste quantities to a user fee for SWS. We do this by calculating the decline in waste that a community can expect from switching to a user fee. Switching from a flat fee for SWS, or from a fee deducted from property taxes, to a $1.00 charge per 32-gallon container would lead to a 15 percent reduction in residential waste in a community that discarded the average quantity for the sample - 2.60 pounds per capita per day. To further indicate the residential response to a user fee we calculate the price elasticity of demand for residential SWS. It turns out to be -0.12 at sample means. Finally, we estimate the welfare gain that could be earned by a community switching to a user charge for SWS. We estimate welfare gains that range from $36,000 per year in a small community with a very low social cost of refuse disposal, to $7,700,000 per year in a community with the opposite characteristics.

The results of our estimation also give the response of residential and commercial waste quantities to changes in such variables as household income, family size and the amount of precipitation. We discuss the impact of changes in these demographic and weather variables in light of results obtained by other researchers.

6.2 RESULTS OF THE GLS ESTIMATION

Table 6.1 presents the coefficient vector estimated by GLS with the complete data set. The table also gives the t-statistics that correspond to each estimate. A t-statistic with an absolute value greater than or equal to two suggests that the coefficient is significantly different from zero at the 5 percent level of significance.

Table 6.1 Estimated Coefficients*

Variable	Coefficient	t-Statistic
DUMMY VARIABLES FOR INTERCEPTS		
Residential Dummy for San Francisco	-57.16	- 3.09
Commercial Dummy for San Francisco	74.42	2.72
Residential Dummy for Hillsborough County	- 9.73	- 2.00
Commercial Dummy for Hillsborough County	81.21	3.75
Residential Dummy for St. Petersburg	-11.84	- 2.05
Residential Dummy for Estherville	- 8.82	- 2.19
Commercial Dummy for Estherville	32.81	4.89
Residential Dummy for Howard County	- 0.66	- 0.23
Combined Residential and Commercial Dummy for Highbridge	- 3.60	- 1.03
Residential Dummy for Bernalillo County	- 0.43	- 0.16
Commercial Dummy for Bernalillo County	8.82	12.29
Residential Dummy for Seattle	-20.97	- 2.90
Residential Dummy for Spokane	-41.73	- 7.20
Commercial Dummy for Spokane	97.42	10.07
RESIDENTIAL SECTOR REGRESSORS		
User Fee for SWS (price per 30- to 32-gallon container)	- 0.40	- 3.99
Average Household Income (in thousands)	0.05	2.87
Mean Temperature (in degrees Fahrenheit)	0.01	9.64
Average Precipitation (in inches)	0.04	5.37
Average Household Size	- 2.75	- 2.66
Age Distribution of the Population (percent of population 18 to 49)	0.11	4.85

Population Density (in thousands)	3.64	3.27
Price Received for Old Newspapers (per short ton)	0.001	0.38
COMMERCIAL SECTOR REGRESSORS		
User Fee for SWS (weekly price per cubic yard for two pick-ups each week)	- 0.23	- 2.58
Mean Temperature (in degrees Fahrenheit)	0.02	4.32
Average Precipitation (in inches)	- 0.04	- 1.19
Population Density (in thousands)	- 4.31	- 2.46
Price Received for Old Corrugated Containers (per short ton)	0.001	0.42

N=600
R^2=0.9207

---

*The dependent variable for the residential equation is measured as pounds of refuse discarded per capita per day. The mean value of this dependent variable for the sample is 2.60. This mean is based on the average pounds per capita per day for communities for which we had data representing residential tonnage.

The dependent variable for the commercial equation is measured as pounds of refuse discarded per employee per day. The mean value of this dependent variable for the sample is 7.50. This mean is based on the pounds per employee per day for the community for which we had data representing commercial tonnage - Bernalillo County.

6.21 The Intercepts

The dummy variable coefficients give the intercepts for the residential and/or commercial models for each of the nine different communities represented by our data set. Interestingly, the residential demand model intercepts are always negative while the commercial model intercepts are always positive.

As expected, the magnitudes of the intercepts vary widely from community to community, because the intercepts reflect the nonrandom differences between communities. For example, open burning of refuse is legal in Estherville but not in any of the other sample communities. Thus, the intercept for Estherville should capture the impact that the legality of open burning has on waste quantities there.

There were too few observations for Highbridge to estimate a separate intercept for its residential and commercial models. Thus, only one dummy variable coefficient is given for Highbridge. It reflects the sum of the intercepts for Highbridge's commercial and residential models.

6.22 The Residential Equation's Demographic and Weather Regressors

Before we consider the impact of a user fee on the demand for residential SWS, let us first consider the impact of changes in certain demographic and weather variables. We shall start with a consideration of the relationship between solid waste and household income.

As reported by Table 6.1, the coefficient for average household income is positive and significant. Its value, 0.05, suggests that when a household's income increases by $10,000, its waste per capita will increase by half a pound per day. This is a fairly large increase when compared to the sample mean value of 2.60 pounds of household refuse discarded per capita per day.

The income elasticity of residential demand for SWS, when calculated at sample means, is 0.41. This suggests that a 10 percent increase in income is associated with roughly a 4 percent increase in discarded waste.

A positive association between household waste and income was not entirely expected. In general, consumption increases with household income, and so one might expect a positive association. On the other hand, as income increases, purchases of starchy and sugary foods and beverages may decline. The latter are purchased by low-income families for their high calorie content and are typically associated with a lot of packaging waste.[1] Exaggerating this effect is the likelihood that a low-income family would purchase such items in small units because of the household's binding budget constraint.

There is another reason we did not entirely expect a positive association between household waste and income. As income increases, people dine in restaurants more often and tend to donate their old clothing to charities rather then discard it.

Recall our conclusion from Chapter 2: The impact of changes in income on the aggregate quantity of waste a household discards is difficult to predict. Other studies have found that certain waste materials, such as green bottles, increase with income while other waste materials, such as textiles, decrease. We find a fairly strong, positive correlation suggesting that there are more waste materials increasing than decreasing with income.

The weather variables are positively correlated with residential waste quantities. This is reasonable since, generally speaking, yard waste should increase during the growing season when there are warmer temperatures and greater precipitation. The temperature coefficient suggests, for example, that for every 10-degree increase in temperature, residents discard a tenth of a pound more waste per day. Besides its correlation with the growing season, the precipitation coefficient is influenced by another factor: All types of absorptive waste are heavier when exposed to rain.

The coefficient for average household size is negative and significant. In other words, there is an inverse relationship between average household size and the per capita demand for SWS. As more individuals are added to a family, the average refuse per family member declines. This suggests that there are indeed economies of scale within the household. Economies of scale exist when financial savings result from sharing among household members. For example, regardless of the number of family members, each household typically has only one yard and only one newspaper subscription. In addition, the larger the household, the lower we expect the packaging waste per capita to be since large households are more apt to purchase 'family sized' food and personal care items. Finally, large households are more likely to hand down clothing, toys and other goods from one family member to another. Thus we depart from other researchers by finding evidence of economies of scale within the household based on the relationship between household size and per capita waste.

The estimated coefficient, -2.75, suggests that the economies are rather substantial. Of course, the average size of a household showed very little change in the communities represented by our data. The average size among the sample communities is 2.39 people, with a sample variance of only 0.05. Thus a decrease of one standard deviation (obtained by taking the square root of 0.05, which gives 0.22) in household size should be considered a moderate decrease. Such a change would increase the quantity of waste per capita per day by half a pound.

Recall that the variable representing the age distribution of the population is the percentage of the population aged 18 to 49. These young adults are expected to generate high refuse quantities simply because they are the greatest consumers of material goods within a society. As Table 6.1 indicates, the coefficient for the age distribution variable is positive and significant, as expected. Its value, 0.11, suggests that if the young adult population of a community increased by 1 percent, that the quantity of refuse discarded there would increase by a tenth of a pound per capita per day.

The sign of the coefficient for the population density variable is positive for the residential equation. If we consider population density as a proxy for urbanization, it is easy to see how its effect on the quantity of refuse could have gone in either direction. Greater urbanization means better access to retail outlets so that goods can be purchased in small quantities with high frequency. This leads to a lot of packaging waste. Urbanization also implies little storage space so that items are necessarily purchased in small increments. Easily replaceable items such as gift boxes and spare automobile parts may be discarded and later repurchased rather than being placed in storage for lengthy periods. These observations suggest that urban populations generate a high volume of waste. Suggesting just the opposite, however, is the small yard size of typical urban households.

To confound matters, the direction of the relationship between population density and the quantity of yard waste is probably not consistent. Urban populations have small yards and little yard waste. Suburban populations have medium to large yards and, typically, a lot of yard waste. Thus, we should expect a negative relationship between yard waste and population density. However, as population density becomes very low, as in rural communities, yards are large enough to maintain a compost pile or yard waste dump. Thus, the demand for SWS for yard waste is probably low in very sparsely populated areas. Overall, the impact of changes in population density on the quantity of refuse set out for disposal was difficult to predict. Our findings suggest that the overriding effects of urbanization are those that lead to higher per capita quantities of residential waste.

The price received for old newspapers should be a good proxy for the price households receive for recycling. At present, 80 percent, by weight, of all refuse recycled is paper, and by far the most commonly recycled paper is newspaper (Franklin Associates, Ltd. 1988, p. 22). We assume that information on the market price being paid for old newspapers reaches households after a lag. Entrepreneurs are the ones who directly respond to price. Often they are a local Boy Scout troop or an elementary school class. Such groups may respond to fluctuations in the price being paid for old newspapers by altering the convenience with which households can recycle newspapers. This suggests that the variable representing the price received for old newspapers be defined as the average of the price received over the past six months. As Table 6.1 indicates, this lagged price has an insignificant impact on the quantity of waste set out by households. Perhaps the speed with which entrepreneurs respond to changes in price varies over time and over communities so that any link between household waste quantities and price is muted.

6.23 The Commercial Equation's Demographic and Weather Regressors

The weather, the density of population and the price of old corrugated containers should affect the quantity of commercial waste discarded. We shall consider the impact of each of these variables. We begin with a discussion of the impact of temperature and rainfall.

We initially included the two weather variables as independent variables in the commercial model for the same reason we included them in the residential model. We expected yard waste to be greater during the growing season and thus to be positively correlated with average temperature and precipitation. As Table 6.1 shows, however, unlike the residential regressors, of the two weather variables, only mean temperature is significant in explaining the quantity of commercial waste per employee. Its coefficient is positive as expected. The insignificant precipitation coefficient is, unexpectedly, negative. Why would the coefficient for the precipitation variable be insignificant? Possibly because yard waste simply is not as large a component of commercial waste as it is of residential waste. This suggests that the mean temperature variable might be significant not only because it correlates with yard waste, but also because it correlates with business activity. Work that must be done out-of-doors typically increases during warm weather. It is reasonable to expect that extra commercial waste will be associated with additional outdoor business activity. Note, however, that such activity increases only when the weather is both dry and warm. Rain would reduce it. This might explain the negative coefficient estimated for the precipitation variable.

The coefficient for population density is negative and significant for the commercial demand model. If we continue to interpret population density as a proxy for urbanization, its coefficient suggests that greater urbanization leads to less commercial waste per employee. The types of businesses that operate in an urban setting can be very different from those in rural areas. For example, the number of service establishments increases with urbanization.[2] These labor-intensive businesses are probably responsible for only a small portion of the commercial waste stream. They do, however, hire many employees. Thus, urbanization is probably associated with a lower quantity of commercial waste per employee. However, because the number of employees increases, urbanization is probably also associated with a higher total quantity of commercial waste.

Let us consider the impact of changes in the price that firms receive for recycling on the quantity of commercial waste. Over 80 percent of recycled waste in the US is paper, and a firm's major contribution to paper recycling comes in the form of corrugated boxes (Franklin Associates, Ltd. 1988, p.

22). Thus, we represent the price that firms receive for recycling as the price received for old corrugated containers. The variable representing this price is defined as an average of lagged prices, in the same manner that the variable representing the price received for old newspapers was, and for similar reasons. We expected the coefficient for this price variable to be negative. The estimated coefficient, however, was not significantly different from zero, perhaps for the same reasons that the estimated coefficient for the price of old newspapers was not.

6.24 The Impact of a Residential User Fee

The regressor of central concern to this study is the residential user fee for SWS. Recall that we have defined it as the average price charged per 30- to 32-gallon container. We did not lag this variable because most municipalities are careful to notify customers of any impending price changes well in advance.

As Table 6.1 shows, the coefficient for the residential user fee variable is negative and significant. This suggests that a positive price for SWS causes the demand for SWS or the quantity of residential waste discarded to decline. Thus, the demand curve drawn in Figure 1.1 does indeed slope downward.

To get an idea of the magnitude of the response to a user fee, we consider three possible user fees that a community might charge. The three range from $0.50 to $2.00 per 32-gallon container and thus encompass the average user fee charged in the sample communities; namely, $0.81.[3] We estimate the decline in waste that would accompany a switch from paying for SWS out of property taxes to charging one of the three user fees. The decline is estimated for a community with the following two characteristics: First, residents there discard the average quantity of waste discarded by the sample and second, the community's population is either 100,000 or 500,000. Table 6.2 summarizes the estimates.

Let us consider a hypothetical community in which residents discard 2.60 pounds of household waste per capita per day, the same as the sample average, and where SWS are financed through property taxes. By how much would a user charge for SWS reduce the quantity of residential waste? Let us assume that the community adopts a new user fee program and charges residents $0.50 per 32-gallon container. The coefficient estimated for the user fee variable suggests that waste would decline by 8 percent to 2.40 pounds per person per day. If the population of the hypothetical community is 100,000, then the annual quantity of waste there will be over

3,500 tons lighter due to the $0.50 user fee. If the population is 500,000, waste will be over 18,000 tons lighter.

Now let us assume that the hypothetical community sets the user charge at $2.00 per 32-gallon container. The user fee coefficient suggests that the quantity of residential waste will decline by 31 percent or by over three quarters of a pound per person per day. If the community has a population of 100,000, this means an annual decline in waste of 14,600 tons. If the population is 500,000, the decline will be 73,000 tons. Table 6.2 summarizes these results and presents the impact of a $1.00 user charge. The latter leads to a decline in waste of 15 percent.

Table 6.2 The Impact of a User Fee

User Fee Per 30- to 32-Gallon Container for Residential SWS ($)	Decline in Waste Per Person Per Day (pounds)	Percent Decline from the Sample Average Quantity of Residential Waste	Annual Decline in Waste with Population = 100,000 (tons)	Annual Decline in Waste with Population = 500,000 (tons)
0.50	0.2	7.7	3,650	18,250
1.00	0.4	15.4	7,300	36,500
2.00	0.8	30.8	14,600	73,000

The overall suggestion is that a user fee can make a substantial contribution to reducing the solid waste stream. This could mean significant savings to a community depending on its spending for such SWS as waste collection and disposal. For the moment, let us consider only a community's spending for solid waste disposal; that is, the tipping fee[4] that it pays. Let us assume that the fee is at a medium level for the late 1980s, say $40.00 per ton of waste. If the hypothetical 100,000-person community paid such a tipping fee, a user charge of $1.00 would reduce its annual disposal bill by almost $300,000. Of course, the larger the population of a community and the higher its disposal and collection costs, the more money it would save. For example, a community with a population of 500,000 that faced a tipping fee of $60.00 per ton would save $2.2 million per year in disposal costs.

A user fee might be desirable even if we disregard the possible money savings. Possibly the most important consequence of a user charge for many communities is that it would slow the rate at which the local landfill was being filled and postpone the need to replace it.

Another way to get an idea of the responsiveness of households to a user charge is to consider the elasticity implied by the user fee coefficient. This elasticity, when calculated at sample means is -0.12.[5] It suggests that waste will decline by about 1 percent in response to a 10 percent increase in the residential user fee. Consider the hypothetical community described above. If officials there increased the user fee from $1.00 to $1.10, waste would decline by 0.03 pounds per person per day. If the community has a population of 100,000, the extra dime per bag would lead to approximately 500 fewer tons of refuse per year. Thus, while the residential elasticity is certainly not large, it is enough that a substantial reduction in waste could follow a very small increase in price.

6.25 The Welfare Gain from a Switch to Residential User Fees

Given that the response of the household to a user fee for SWS is significant, how large is the welfare gain to society caused by a switch to user fees? We will present several estimates of this welfare gain. The first is based on Seattle data. Seattle is a community that has suffered from a scarcity of disposal sites and thus is representative of communities that face a high cost of disposal. However, even Seattle does not consider public costs such as the expected cost of surface water contamination from a landfill when pricing SWS. Thus, we calculate another estimate of the welfare gain and take such public costs into account. Unlike Seattle, some US communities are not experiencing a severe shortage of landfill space so we calculate an estimate of the welfare gain based on the average cost of disposal across the country as reported by the EPA. Finally, we estimate the welfare gain from a switch to user fees for a community that has a very low cost of disposal. Table 6.3 presents the four estimates.

Seattle is the only city in the sample that sets its residential user fee so that the private costs of SWS are recouped. For this reason, Seattle's user fee can be considered an estimate of the average private cost of residential SWS in Seattle. To calculate the welfare gain from a switch to user fees requires an estimate of the marginal *social* cost of SWS.[6] Henceforth we will take the marginal private cost as a lower bound of the marginal social cost of SWS. This is reasonable since social costs include private costs -

those explicit costs whose dollar values are easy to see - as well as public costs such as the cost of the possibility that a community's ground water will be contaminated by a landfill. For the sake of simplicity, we assume that SWS operations have constant returns to scale so that the marginal cost of SWS equals its average cost.

For the reader's convenience, a revised Figure 1.1 appears below as Figure 6.1. The residential user fee for SWS in Seattle in 1988 was $1.71 per 32-

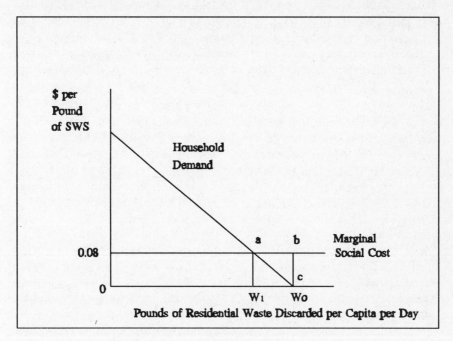

Figure 6.1 The Efficiency Gain from User Fees

gallon container. According to research by the Seattle Solid Waste Utility, a 32-gallon refuse container holds an average of 20 to 22 pounds of household waste (Skumatz 1990, p. 12). Thus, the 1988 user fee per pound, which we take as the social cost of SWS per pound, was approximately $0.08. To obtain the value of the welfare gain triangle, abc, in Figure 6.1, we also need estimates of the quantities of waste represented by points W_1 and W_0. These points are just the predicted values of residential pounds discarded in Seattle per capita per day given a user fee of $0.08 per pound (point W_1) and given a user fee of zero per pound (point W_0). The value of point W_1 is calculated by combining the coefficient vector given in Table 6.1 with the average 1988 values of Seattle's independent variables, which include a value for the residential user fee of $1.71 per 32-gallon container. This calculation yields a value for point W_1 equal to 1.67. The value of W_0 is calculated in a similar manner, except that the user fee per 32-gallon container is set equal to zero. This point is estimated to have a value of 2.35. Finally, we obtain an estimate of the annual welfare gain if Seattle switched from a flat fee for SWS to a fee of $0.08 per pound, assuming that the true social cost of SWS was in fact $0.08 per pound. This annual gain equals approximately $5,000,000. To put this estimate into perspective, consider that Seattle's total operating budget for SWS in 1988 was approximately $125,000,000. The welfare gain is 4 percent of this total.

Seattle is representative of US cities experiencing a crisis in municipal SWS. Seattle's cost of SWS is probably higher than in many US communities. To get a rough idea of the welfare gain that could be experienced by an average US community we used data from a 1986 EPA survey of 1,250 municipal landfill facilities throughout the US and five US territories. The results of that survey suggest that the average operating costs of a landfill are approximately $50.00 per ton of refuse (US EPA 1987, p. G15). Several studies report that landfill costs make up between 20 and 40 percent of total SWS costs.[7] Taking into account that relative to other SWS costs, landfill costs have risen at a rapid pace during the late 1980s, we will make the conservative assumption that landfill costs are 40 percent of total SWS costs.

These assumptions lead to an estimate of the private cost per ton of SWS in an average US community of $125.00. This, in turn, suggests a cost of $0.06 per pound of SWS. Continuing to assume that the private cost of SWS is an estimate of the lower bound of its social cost, an efficient user fee should equal $0.06 per pound, or $1.31 per 32-gallon container.

To get an idea of the average community's welfare gain, we multiplied the residential user fee coefficient given in Table 6.1 by $1.31. The product suggests that households will respond to an increase in the user fee from zero to $1.31, all other variables constant, by reducing the pounds of waste discarded per person per day by one half pound. This suggests that the welfare gain from switching to an optimal user fee for SWS in an average US community, per person per day is $0.016. Thus, even a rather small community with a population of only 100,000 would enjoy a welfare gain of approximately $600,000 annually. A larger community with a population of 500,000 would experience a $3,000,000 gain.

The three welfare gain estimates given above are summarized in the first rows of Table 6.3. The table presents two additional pairs of estimates. One pair is calculated using a value for the social cost of residential SWS that is quite high; namely, $0.10 per pound or $2.10 per 32-gallon container. This estimate is based on the assumption that the private costs of disposal do not reflect all of the social costs of SWS. It seems reasonable to expect that there are at least some public costs of SWS that are not reflected by the costs paid by the municipality for SWS; for example, the expected cost of possible ground or surface water contamination from a landfill. Imposing a user fee that reflects a $0.10 per pound cost suggests that a community with 100,000 residents will enjoy an annual welfare gain of $1,500,000; a community with 500,000 residents will gain $7,700,000. Of course these results only hold if everything, besides the user fee, that is expected to affect the demand for SWS remains constant.

Table 6.3 Estimates of the Welfare Gains Attributable to User Fees for SWS			
Region	Optimal User Fee[a] ($)	Decline in Waste[b]	Welfare Gain[c] ($)
Seattle (1988)	1.71	0.68	5,000,000
US Community population=100,000 population=500,000	1.31	0.52	600,000 3,000,000
US Community population=100,000 population=500,000	2.10	0.84	1,500,000 7,700,000
US Community population=100,000 population=500,000	0.32	0.13	36,000 180,000

a. This column gives the optimal user fee per 32-gallon container, which is equal to the assumed marginal social cost of SWS.
b. This column gives the decrease in the pounds of residential waste per person per day if the residential user fee were to increase from zero to its optimal level.
c. This column gives the approximate annual welfare gain associated with a residential user fee equal to the marginal social cost of SWS.

As a final estimate of the change in welfare due to a switch to an efficient user fee, consider a very low estimate of the social cost of SWS. The EPA's 1986 survey estimates that the operating costs of the average US landfill are $50.00 per ton. This cost estimate is high relative to that offered by Glebs (1988). Glebs estimates that a landfill located in an upper-Midwestern state and designed to meet the new relatively strict EPA Subtitle D requirements will have a total private cost of between $12.02 and $27.70 per ton of waste accepted. Assuming once again that the private cost of SWS represents the lower bound of its social cost, we take the lower limit of Glebs' range of estimates, namely $12.02, as a final estimate of the social cost of SWS. We further assume that this cost makes up 40 percent of the total costs associated with SWS. This suggests that the social cost of disposing of refuse is a very low $0.015 per pound, or $0.32 per 32-gallon container. Such conservative data suggest that the annual gain associated with a switch to an optimal user fee is $36,000 in a community with a population of 100,000 and $180,000 in a community with a population of 500,000.

6.26 The Impact of a Commercial User Fee

We have defined the commercial user fee variable as the average weekly price charged per cubic yard of dumpster capacity for two pick-ups per week. Again, we have not lagged this price variable because municipalities typically notify customers in advance of impending price changes. Table 6.1 reveals that the user fee coefficient for the firm is negative and significant. This coefficient and the sample mean values of the relevant variables yield an estimate of the elasticity of commercial demand for SWS of -0.29.[8] Thus our results suggest that the response of commercial establishments to changes in the price of SWS is greater than the response of households. The average user fee charged to businesses in our sample was $9.30. Consider the impact of an increase in this price on the average quantity of commercial waste, 7.5 pounds per employee per day. If the user fee rose to $10.00, the pounds per employee per day should fall to 7.34. This suggests that the quantity of commercial waste discarded in a year within a community would fall by 58 pounds for each person working in the community. This response is large enough to suggest that one way to prolong the life of a municipal landfill is to increase the commercial user fee for SWS.

6.27 Comments Regarding the Reduction in Waste from User Fees

The model does not reveal whether households and businesses reduce their waste by recycling more, by buying goods that generate less waste, or by illegally discarding waste. Only the latter is undesirable and conversations with municipal authorities suggest that illegal dumping does happen immediately following imposition of residential user charges. Even so, in the long run, these authorities have found the benefits of a user fee program to outweigh the costs imposed by illegal dumping. To deter the illegal response they suggest rapid and consistent enforcement of anti-dumping laws. With such enforcement, illegal dumping eventually declines. New user fee communities do, however, seem to go through an adjustment period while residents get used to the idea of paying a user fee for something that previously seemed free.

Recall that municipal haulers operate in all of the sample communities. In fact, in all but one community the only service available to residents and businesses was municipal - there were no competing private refuse haulers. This is unlike many communities where residents and especially businesses might have a choice of haulers. Consider a community where businesses have such a choice and assume that officials there increase the price of municipal SWS. The resulting decline in commercial waste may be due to a fourth factor not mentioned above; that is, a transfer of waste from the municipal hauler and the local landfill. to a private hauler and possibly a more distant disposal site. This fourth possibility could reduce commercial waste beyond that suggested by our model.

6.28 Conclusion

The estimates presented in Table 6.3 measure the welfare gains that will be experienced by individual communities. The welfare gain to the entire US from widespread imposition of user fees for SWS should be considerably larger than any one of the estimates in Table 6.2. Given that the administrative costs of a user fee program are comparable to those of a flat fee program or to a policy of deducting a fee for SWS from property taxes, and assuming that there would not be a sizable increase in illegal disposal practices in response to imposition of a user fee, our calculations suggest that there are substantial gains awaiting municipalities willing to charge a user fee for SWS.

NOTES

1. For a more in-depth discussion, see Section 2.41 of Chapter 2.
2. See, for example, Bish and Nourse (1975, p. 41) and Prodromidis and Lianos (1973, p. 100).
3. This average was calculated based only on data from the communities that charged user fees.
4. A tipping fee is the fee paid by haulers, usually per ton of waste, to deposit refuse at a disposal site or transfer station.
5. The sample mean value of the pounds of residential waste per capita per day was determined as the average of the tonnage data that reflect only residential waste quantities. The sample mean value of the residential user fee was calculated using data only for those communities that charged positive user fees.
6. For a discussion of the reasons why, please see Chapter 1.
7. See, for example, Beck (1987), Neal and Schubel (1987) and Organization for Economic Co-Operation and Development (1981, p. 14).
8. The sample mean value of the commercial pounds per employee per day is the average of the quantity data that reflects only commercial waste quantities.

7. Tests and Respecifications of the Empirical Model

7.1 AN OVERVIEW

There are a number of assumptions underlying the demand model presented in Chapter 4. One is that the true model relates to monthly, and not yearly or quarterly, data adjustments. In order to check the sensitivity of the results to the monthly specification, we specify an annual model. This annual model and the coefficient vector estimated for it are described in detail below. As we shall see, the annual model's coefficient vector is not unlike the one for the monthly model.

Two additional assumptions that underlie the demand model of Chapter 4 are that the prices charged for SWS are exogenous and that the slope coefficients of the model are the same across communities. We have tested each of these assumptions and present our results below.

7.2 AN ANNUAL VERSION OF THE GENERALIZED LEAST SQUARES (GLS) MODEL

The model considered in this text is a monthly one. However, the data for many of the independent variables are yearly (see Table 5.4). To get monthly values from yearly data, we simply repeated the yearly values 12 times. Thus the monthly model contains some errors of measurement.

For comparison purposes, we formulated and estimated an annual model. If the results of the monthly and annual models are qualitatively and quantitatively similar, we will assume that the measurement errors involved in the monthly model are not 'serious.'

To complicate matters, we expect the annual model to contain its own errors of measurement. If the basic formulation of the demand model for SWS is monthly - that is, the data are adjusted in monthly intervals - then an annual model formulated entirely in terms of annual variables will contain errors of measurement. The reason is that the model is nonlinear. To see the issue involved, consider one of the monthly terms in equation

(5.1), namely,

$$\frac{X_{it1}^R}{CPI_{it}} \tag{7.1}$$

and assume that the monthly specification is correct. Then, in a correctly formulated annual model, the variable that would correspond to (7.1) would be a weighted average of that ratio over the entire year. For example, one annual value of such a variable would be

$$\frac{31\left(\dfrac{X_{i,1,1}^R}{CPI_{i,1}}\right) + 28\left(\dfrac{X_{i,2,1}^R}{CPI_{i,2}}\right) + \ldots + 31\left(\dfrac{X_{i,12,1}^R}{CPI_{i,12}}\right)}{365}. \tag{7.2}$$

If, however, monthly data are not available, the term in (7.2) cannot be calculated. Instead, the expression in (7.2) might be approximated by a term such as

$$\frac{[31(X_{i,1,1}^R) + 28(X_{i,2,1}^R) + \ldots + 31(X_{i,12,1}^R)]/365}{[181(CPI_{i,1}) + 184(CPI_{i,7})]/365}. \tag{7.3}$$

In formulating our annual model we relied on approximations similar to that suggested by (7.3). Thus, nonlinearities in the monthly model have led to measurement errors for the annual model.

In specifying the annual model we took pains to ensure that the magnitudes involved were of the same dimensions as those in the monthly model. Without such care, a comparison of the two models would not be very meaningful. For example, the dependent variable in the annual residential model is pounds of waste discarded per capita per day, just as it is in the monthly model. The difference is that in the annual model the dependent variable for a given year is calculated as the average over that entire year; in the monthly model, the dependent variable for a certain month is simply the average for that month.

The error structure of the annual model is similar to that of the monthly model. Consider the disturbance term in the annual residential equation:

$$\left(\frac{31}{365}\right)\epsilon_{it}^R + \left(\frac{28}{365}\right)\epsilon_{it}^R + \ldots + \left(\frac{31}{365}\right)\epsilon_{it}^R \tag{7.4}$$

This term is equal to a weighted average of twelve monthly disturbance terms. The reason is that each annual equation is the sum of twelve monthly equations. Thus, the only difference between the variance-covariance matrix of the annual model and that of the monthly model is that all the variance terms for the annual model are larger than the corresponding monthly ones by a fixed proportion. The fixed proportion equals the sum of the squared weights that would be applied to the monthly observations in a correctly formulated annual model - see (7.2). Thus, the method of estimation used for our annual model is that suggested by its error structure, namely, generalized least squares (GLS).

The coefficient estimates for the annual model and their corresponding t-statistics are presented in Table 7.1.

A comparison of the coefficient estimates of the annual model with those of the monthly model (see Table 6.1) suggests that there are no dramatic differences. With four exceptions, the coefficients in the annual model have the same sign as the corresponding coefficients in the monthly model. The magnitudes of the coefficients are reasonably similar as well.

One of the four exceptions is the intercept estimated for the commercial equation for Bernalillo County. This coefficient is negative and insignificant in the annual model but positive and significant in the monthly model.

Two of the four exceptions are the coefficients for the price of old newspapers and the price of old corrugated containers. However, since the coefficients for these price variables are insignificant anyway, their signs are probably not very meaningful.

The coefficient for the household income variable was positive and significant in the monthly model but negative in the annual one. Recall that the sign of the household income coefficient is difficult to predict. The reason is that while we expect consumption in general to increase with income - indicating that the demand for SWS would increase - we also expect some types of waste, such as textiles, to decrease with income. The results of the annual model cast doubt on our earlier assertion that a household's income is positively correlated with its demand for SWS.

Overall, the coefficient vector estimated for the annual model is reasonably close to the one estimated for the monthly model. This suggests that the errors of measurement involved in the monthly model may not be overly important in determining the results; in other words, the results may not be primarily determined by a (less than perfect) monthly specification.

Table 7.1 Coefficient Estimates for the Annual Model*

Variable	Coefficient	t-Statistic
DUMMY VARIABLES FOR INTERCEPTS		
Residential Dummy for San Francisco	-64.82	- 1.43
Commercial Dummy for San Francisco	65.14	0.92
Residential Dummy for Hillsborough County	-22.03	- 1.31
Commercial Dummy for Hillsborough County	116.97	1.53
Residential Dummy for St. Petersburg	-16.91	- 1.15
Residential Dummy for Estherville	-10.91	- 1.03
Commercial Dummy for Estherville	17.32	0.75
Residential Dummy for Howard County	- 3.61	- 0.44
Combined Residential and Commercial Dummy for Highbridge	- 8.23	- 0.86
Residential Dummy for Bernalillo County	- 4.04	- 0.51
Commercial Dummy for Bernalillo County	- 3.15	- 0.36
Residential Dummy for Seattle	-25.50	- 1.44
Residential Dummy for Spokane	-47.06	- 3.71
Commercial Dummy for Spokane	91.61	4.53
RESIDENTIAL SECTOR REGRESSORS		
User Fee for SWS (price per 30- to 32-gallon container)	- 0.28	- 1.11
Average Household Income (in thousands)	- 0.001	- 0.03
Mean Temperature (in degrees Fahrenheit)	0.04	0.59
Average Precipitation (in inches)	0.04	0.37
Average Household Size	- 1.09	- 0.41
Age Distribution of the Population (percent of population 18 to 49)	0.09	1.76

Population Density (in thousands)	3.96	1.48
Price Received for Old Newspapers (per short ton)	- 0.001	- 0.15
COMMERCIAL SECTOR REGRESSORS		
User Fee for SWS (weekly price per cubic yard for two pick-ups each week)	- 0.41	- 1.90
Mean Temperature (in degrees Fahrenheit)	0.25	1.64
Average Precipitation (in inches)	- 0.07	- 0.25
Population Density (in thousands)	- 4.41	- 1.00
Price Received for Old Corrugated Containers (per short ton)	- 0.004	- 0.81

N=49
R^2=0.99773

---

*The dependent variable for the residential equation is measured as pounds of refuse discarded per capita per day. The mean value of this dependent variable for the sample is 2.62. This mean is based on the average pounds per capita per day for communities for which we had data representing residential tonnage.

The dependent variable for the commercial equation is measured as pounds of refuse discarded per employee per day. The mean value of this dependent variable for the sample is 7.54. This mean is based on the data for Bernalillo County, the only community for which we had separate data related to commercial tonnage.

7.3 A HAUSMAN TEST FOR ENDOGENEITY OF THE USER FEE VARIABLES

The models considered in this paper so far have been built on the assumption that the user fee variables, X_{it1}^R and X_{it1}^C in (5.1) through (5.3), are exogenous. In actuality the user charge for SWS might be simultaneously determined with the demand for SWS. In particular, the waste collection agency might raise the user fee in response to high quantities of discarded waste.

There are reasons to suspect, however, that the causal link running from quantities of waste to user fees is a weak one. Recall that the waste collection agencies represented by our data are all municipally controlled or operated. We suspect that, as a result, changes in user fees are very closely linked to demonstrable changes in collection and disposal costs, such as diesel fuel price changes, rather than incremental changes in the demand for services.

The response of user fees to at least one demonstrable change in disposal costs could suggest that the user fees are endogenous. As a disposal site approaches capacity, there is sometimes an accompanying increase in tipping fees. Consider the following scenario. As waste quantities go up, the local disposal site approaches capacity. To prolong the life of the disposal site and avoid the high fixed costs of opening a new one, the disposal agency raises its tipping fee. The collection agency, possibly a separate institution, then raises its residential and commercial user fees to help pay for the higher disposal costs. This scenario explains how, during the final operating years of a disposal site, high quantities of waste could lead to higher user fees. Note that there would probably be a time lag between the increased quantities of waste and the increased user fees, since the user fees would likely change only after tipping fees had changed.

Our suspicion that there is a weak causal link from quantities of waste to user fees, and the fact that the aforementioned scenario involved a lag, led us to assume that the user fee variables are predetermined. To test this assumption we performed a Hausman test for the endogeneity of the two user fee variables. Using the notation developed for (5.1) through (5.3) the null and alternative hypotheses are

$$H_0: \quad X_{it1}^R, X_{it2}^R, ..., X_{it8}^R,$$
$$X_{it1}^C, X_{it2}^C, ..., X_{it5}^C \text{ are predetermined,}$$

$$i = 1,...,9,$$
$$t = 1,...,T_i ;$$

$$H_1: \quad X_{it1}^R, X_{it1}^C \text{ are endogenous,}$$
$$X_{it2}^R, X_{it3}^R, ..., X_{it8}^R,$$
$$X_{it2}^C, X_{it3}^C, ..., X_{it5}^C \text{ are predetermined,}$$

$$i = 1,...,9,$$
$$t = 1,...,T_i.$$

Under the null hypothesis, the model remains as specified in Chapter 4. Under the alternative hypothesis, the model becomes a system of equations.

If the alternative hypothesis is true, the GLS estimator is not consistent. An estimator that would be consistent under both the null and alternative hypotheses is a two-stage least squares (2SLS) estimator. Let

β^G = the GLS estimator of the coefficient vector under the null hypothesis, and

β^{2S} = the 2SLS estimator of the coefficient vector under the alternative hypothesis, which accounts for the error structure of the disturbance terms.

Assuming a type 1 error of 0.05, the Hausman test uses the result that

$$H = T\, g'V^{-1}g \;\rightarrow\; \chi^2_k \qquad (7.5)$$

where

$g\quad = \beta^G - \beta^{2S}$,

Plim V = the difference between the asymptotic variance-covariance matrices of the coefficient vectors estimated by GLS and 2SLS,

$k\quad$ = the number of exogenous variables in the model under the null hypothesis.

We reject the null hypothesis if H is greater than $X^2_k(0.95)$ where

$$P[\chi^2_k \le \chi^2_k(0.95)] = 0.95. \qquad (7.6)$$

To do 2SLS we used several instrumental variables; namely,

X^N_{it1} = the average wholesale price per gallon of diesel fuel in community i at time t, deflated by the national producer price index (PPI);

X^N_{it2} = the average annual wages earned per worker employed in 'refuse systems' in community i's state in period t, deflated by the national PPI;

X^R_{it9} = the average pounds of residential waste discarded per capita per day in community i over the six-month period prior to month t and;

X^C_{it6} = the average pounds of commercial waste discarded per employee per day in community i over the six-month period prior to month t.

Unfortunately, the earliest data available for regional wholesale diesel fuel prices were for January 1983, so that became the earliest data point that we used for the Hausman test.

We have only 12 monthly observations for community six, Highbridge, which gave us too few degrees of freedom to include the data for this community in the 2SLS estimation. Thus we dropped the Highbridge data from the analysis for the Hausman test.

One final caveat, concerning the construction of the two lagged variables, X_{it9}^{R} and X_{it6}^{C}. These variables represent the recent level of demand for residential and commercial SWS, respectively. We actually have data that represent the demand for residential SWS for only four communities; in particular, communities three, five, seven and eight. We have data that represent the demand for commercial SWS for only one community; namely, community seven. For all the other communities our data represent the sum of the demand for residential and commercial SWS. For the latter communities we had to use the average quantity of combined residential and commercial waste per capita for the six-month period prior to the current month as a proxy for both X_{it9}^{R} and X_{it6}^{C}.

We estimated the coefficient vector by 2SLS in the following manner. Before either stage, we corrected the model for the heteroskedasticity described in Chapter 4. During the first stage, we determined the calculated value for the residential (commercial) user fee as a function of a constant term and all the exogenous residential (commercial) variables, including the instrumental variables. In the second stage we replaced the original user fee variables with the estimates of them obtained in the first stage. We then proceeded with an ordinary least squares (OLS) estimation procedure.

In summary, for the Hausman test we have estimated the coefficient vector by GLS and by 2SLS. Both estimations are based on truncated versions of the original data set. In particular, the earliest data we used were for January 1983, and we dropped the data for Highbridge. The GLS estimator was obtained as discussed in Chapter 4. The 2SLS estimator was obtained as described immediately above.

Inserting the results of the 2SLS and the simple GLS estimations into (7.5) gives $H = 0.24321$. The critical value of the chi-square distribution at a 5 percent significance level and with $k = 26$ is 15.379 (Johnston 1984, p. 549). Thus we accept the null hypothesis that the user fee variables are exogenous.

This finding supports our initial belief that the causal link from waste quantities to user fees is a weak one. Alternatively, the finding supports the idea that there is a time lapse between a change in the quantity of waste and the resulting change in the user fee for SWS.

7.4 A TEST FOR CONSTANT SLOPE COEFFICIENTS

When specifying the demand equations, we assumed that the intercepts vary from community to community. We also implicitly assumed that the slope coefficients do not. In this section we describe a test of the latter assumption.

We were unable to apply this test to all the communities in the data set because the test requires a separate coefficient vector to be estimated for each community. Communities two, four, five, six, and nine simply had too few observations to permit estimation of the complete coefficient vector. Thus, we apply this test to a truncated version of the data set.

The model in which the slopes are assumed to be the same across communities can be expressed as

$$y_S = X_S \gamma_S + U_S,$$ (7.7)

where y_S, X_S, γ_S and U_S are similar to y, X, γ, and U in (4.17) except that the former only relate to communities one, three, seven and eight. The model in which the slopes are not restricted can be expressed as

$$y_S = X_U \gamma_S^2 + U_S,$$

where (7.8)

$$\gamma_S^{2\prime} = (a_1^R, a_1^C, B_1^{R\prime}, B_1^{C\prime}, a_3^R, B_3^{R\prime}, a_7^R, B_7^{R\prime}, a_7^C, B_7^{C\prime}, a_8^R, B_8^{R\prime})$$

and where X_U is X_S rearranged to correspond to γ_S^2.

We specify the null hypothesis as

$$H_o: \quad B_1^R = B_3^R = B_7^R = B_8^R,$$ (7.9)

$$B_1^C = B_7^C,$$

and the alternative, H_1 as the negative of the null. To test the null against its alternative, we calculate the following chi-square statistic:

$$\chi_q^2 = \frac{(ESS_R - ESS_U)(T-K)}{ESS_U}$$ (7.10)

where

ESS_R = the error sum of squares for the restricted model;
ESS_U = the error sum of squares for the unrestricted model;
q = the number of restrictions;
T = the total number of observations;
K = the number of parameters estimated for the unrestricted model.

Taking the type one error to be 0.05, we reject the null hypothesis if $X_q^2 \geq X_q^2(0.95)$ where $X_q^2(0.95)$ is the critical value given by the chi-square distribution with q degrees of freedom.

To calculate the error sum of squares for the unrestricted model, we first corrected that model for heteroskedasticity. Each community's coefficient vector was then estimated by OLS. The unrestricted error sum of squares was obtained by summing the error sums of squares of the separate demand equations. This led to a value of 22.63 for the ESS_U.

To calculate the error sum of squares for the restricted model, we again corrected the model for heteroskedasticity. We then pooled the data for the four communities and estimated a single coefficient vector by GLS. The sum of squared residuals for the restricted model was 30.87.

Inserting our calculations into the chi-square statistic in (7.10) gave a X_{27}^2 value of 132.19. The value of $X_{27}^2(0.95)$ is 16.151. Thus, we reject the null hypothesis that the slope coefficient vector is constant over communities.

The coefficient vectors estimated separately for each community are summarized in Table 7.2.

As suggested by Table 7.2, the majority of coefficient estimates for individual communities are not significant. Of those that are significant, only two are substantially different than the coefficient estimates for the pooled data given in Table 6.1. In particular, the coefficient estimated for population density in the Seattle residential equation is equal to -2.24 whereas that for the pooled data set is 3.64. The coefficient estimated for average household income in the Bernalillo County equation is equal to -0.06 whereas the one for the pooled data set is 0.05. Recall that during the discussions of both population density and household income we suggested that there are good reasons to expect that the impact of these variables on the quantity of residential waste could be either positive or negative.

Given the nature of the data we pooled, we might have expected the coefficient estimates for individual communities to be different. As discussed in detail in Chapter 5, our data come from many independent sources and there are inconsistencies in the types of waste that it represents. The institutions that one might expect to affect waste quantities are also quite different from community to community. These large   differences

Table 7.2 Coefficient Estimates for Select Communities				
Variable	San Francisco (1)[a]	St. Petersburg (3)	Bernalillo County (7)	Seattle (8)
RESIDENTIAL SECTOR REGRESSORS				
User Fee for SWS	-0.42	NA[b]	NA[b]	-0.10
Household Income	-0.01	0.20	-0.06[c]	0.01
Mean Temperature	0.13	0.03[c]	0.01[c]	0.01[c]
Average Precipitation	-0.52	0.05[c]	0.01[c]	0.01
Household Size	-0.40	-2.53	6.37	-2.68[c]
Age Distribution	-0.01	0.02	-0.15	0.06
Population Density	-0.36	7.56	11.35	-2.24[c]
Price for Old Newspapers	0.001	0.001	0.008[c]	0.002

COMMERCIAL SECTOR REGRESSORS				
User Fee for SWS	0.01		-0.28	
Mean Temperature	-0.16		0.02[c]	
Average Precipitation	0.84		0.13	
Population Density	1.23		-13.47	
Price for Old Corrugated Containers	0.003		-0.004	
	n=99	n=102	n=77, 75[d]	n=102

a. The numbers in parentheses refer to the community numbers.
b. The price coefficient is not estimable since the user fee for SWS is always equal to zero.
c. This coefficient is significant at the 5 percent level.
d. Recall that separate data for residential and commercial quantities of waste are available for Bernalillo County. Thus, the residential and commercial equations for Bernalillo County have been estimated separately. The former equation was estimated with 77 observations, the latter with 75.

suggest that perhaps the data for the nine communities should not be pooled at all.

As Table 7.2 suggests, however, the data lose almost all value when kept separate for the individual communities. Very few coefficient estimates are significant. In other words, the results are very uncertain when the data are not pooled. There are, perhaps, too few observations corresponding to any one community to produce meaningful and significant results.

Pooling the data for the nine communities increases the number of observations substantially. Perhaps the bias introduced by pooling is more than offset by the benefit of having plentiful observations. An increase in observations always reduces the variances of a model's coefficients. Thus, the mean squared errors corresponding to the pooled data model might be less than those corresponding to the model that does not involve pooling.

Clearly, further research is needed. This research might consider the pooled model but in terms of a more high-powered econometric method. One framework in which such a method might be developed is that of random coefficients. In such a framework, one can pool data even though the coefficients differ over the units of observation.

A possibility for the model that we do not test is that the intercepts are equal over the communities. The inconsistencies in the waste tonnage data described in Chapter 5 were great enough to suggest that the intercepts vary among communities.

8. Forecasting Waste Quantities

8.1 INTRODUCTION

The demand models we have estimated may be used to forecast the quantity of waste discarded by a specific community. Two categories of waste may be forecast - residential or commercial. In both cases, several pieces of information are needed. One is data that represent the quantity of residential and/or commercial waste that was discarded in a previous period. For example, the best forecast of residential waste will be produced with past data that represent a residential waste quantity.

In many communities, the only data available will represent the sum of commercial and residential waste. When this is the case, a reasonable estimate of the percentage of the waste stream in that community that is residential (commercial) will provide a second best starting point for producing a forecast of the residential (commercial) waste quantity. If such an estimate is unavailable, the only forecast that may be calculated is of the combined quantity of commercial and residential waste.

Besides the data related to waste quantities, additional data are needed to produce a forecast. Estimates are needed of the previous and future values of certain independent variables. The precise independent variables for which information is needed depends on which category of waste is being forecast - residential, commercial or combined.

This chapter begins by outlining a procedure for forecasting the quantity of residential waste discarded by a specific community. The procedure may also be used to calculate the expected impact of a residential user fee for SWS. Its impact can be estimated simply by comparing two forecasts: one based on the assumption that residents are not charged a user fee and another based on the assumption that they are. Besides the impact of a user fee, a community might be interested in the monetary value of a user fee program. One measure of this value is the welfare gain that a community will enjoy from the program's implementation. Instructions for estimating such a welfare gain are given below. Instructions are also given for forecasting the quantity of commercial waste and the quantity of combined waste. The chapter concludes with some remarks regarding the accuracy of a forecast.

8.2 FORECASTING THE QUANTITY OF RESIDENTIAL WASTE

To forecast the quantity of residential waste discarded by a community, an estimate or, preferably, an actual value of the quantity of residential waste discarded by the community in a previous period is needed. The quantity should be measured as the weight rather than the volume of residential waste. Estimates of the data items listed in Table 8.1 are also needed for the previous period as well as for the future (forecast) period.

Table 8.1 Data Required for a Community to Forecast Its Residential Waste Quantity

---

1. User fee for residential SWS (the price in dollars per 30- to 32-gallon container)

2. Average annual household income

3. Mean temperature in degrees Fahrenheit

4. Total monthly precipitation in inches

5. Average household size

6. Percentage of the population aged 18 to 49

7. Population density (population of the community divided by the square miles of land in the community)

8. Regional price paid for old newspapers per short ton

9. Population

10. Consumer price index divided by 100

Each item above may be the average for a given year or may be the average related to a particular month, except for item 4. It should be the monthly total rather than the average when monthly data are being used.

To illustrate how to use the information in Table 8.1 let us consider a hypothetical community, Wasteville. Wasteville solid waste officials know that the total amount of residential waste discarded there in 1990 was 5,000 tons. The population then was 11,000, so the quantity of waste discarded per capita per day was 2.49 pounds. Wasteville officials are interested in predicting the residential waste quantity for 1995. They expect it to be higher then, given no change in their current policy of paying for SWS out of revenues collected from property taxes. The officials are curious about the impact of a user fee for SWS. They would like to know how much a user charge of $1.00 per 32-gallon container will discourage growth of the waste stream. Thus, Wasteville officials are in fact interested in two forecasts for 1995: one based on the assumption that residents are charged zero per bag of SWS to represent a continuation of their current policy, and a second based on the assumption that residents are charged a user fee of $1.00 per bag. We will work on the former forecast in steps one through six. Then we will discuss the latter forecast and the difference that a user fee might make.

Wasteville officials have gathered actual data for 1990 from the Census Bureau and other agencies for the items listed in Table 8.1. The Census Bureau was able to provide the annual average for most items rather than the averages for any given month. Wasteville officials have also succeeded in obtaining estimates of the future 1995 values for each data item in the table. They received some of these estimates from a local economic development office and some from a local electric utility. The results of their data gathering efforts are given in Table 8.2. The reference numbers that appear there correspond to the reference numbers introduced by Table 8.1.

In general, once the appropriate data are in hand, the following steps may be followed to forecast a waste quantity.

Step One. Deflate all the dollar items by the consumer price index; that is, divide items 1, 2 and 8 by item 10. The revised items for Wasteville are:

	1990	1995
1'.	$0	$0
2'.	$14,050	$15,323
8'.	$1.65	$1.61

Table 8.2 Wasteville Data for the Residential Forecast

	1990	1995
1.	$0	$0
2.	$17,000	$19,000
3.	60	60
4.	2.5	2.5
5.	2.69	2.58
6.	55.2	54.9
7.	2,700	2,823
8.	$2.00	$2.00
9.	11,000	11,500
10.	1.21	1.24

Step Two. Divide the household income and population density data by 1,000. For Wasteville, this leads to:

	1990	1995
2".	$14.05	$15.323
7'.	2.7	2.823

Step Three. For the previous and the future time periods, multiply data items 1 through 8 by the corresponding coefficient estimated for the residential equation. If the data has been revised by steps one and two, use the revised data. The coefficients appear in Table 6.1 but, for the reader's convenience, are repeated below and associated with the data item reference numbers given by Table 8.1.

	Coefficient
1.	- 0.40
2.	0.05
3.	0.01
4.	0.04
5.	- 2.75
6.	0.11
7.	3.64
8.	0.001

For Wasteville, we multiply each data item, using revised data when appropriate, by its corresponding coefficient to obtain the following products.

	1990 Product	1995 Product
1.	$0	$0
2.	$0.7025	$0.7662
3.	0.6	0.6
4.	0.1	0.1
5.	- 7.3975	- 7.095
6.	6.072	6.039

7.   9.828   10.2757

8. $0.0017 $0.0016

Step Four. Sum the products calculated in step three for each time period; that is, add the product associated with data item 1 to the product associated with data item 2 and continue through the product associated with data item 8.

For Wasteville the sum is 9.9067 for 1990 and 10.6875 for 1995.

Step Five. Consider only the sum of the products for the previous period. This sum is necessary to estimate an intercept for the community. The intercept reflects all the unique characteristics of a community that affect the quantity of waste discarded there. For example, the intercept will reflect the laws and rules in a community that govern which types of refuse can be placed at the curb. It also will reflect whether a community has a strong collective environmental conscience. These unique characteristics are not represented by the independent variables, so their effects are not isolated. The lump sum effect of all such characteristics, however, is represented by the intercept.

Subtract the sum of the products for the previous period from the quantity of residential waste discarded per capita per day then. The difference is an estimate of the intercept; albeit a poor one. An easy method - but one that requires data for more than one previous period - for vastly improving this estimate is given below.

Recall that in 1990 the quantity of residential waste discarded per capita per day in Wasteville was 2.49. The latter value minus the 1990 sum calculated in step five, 9.9067, gives -7.4167. This is an estimate of the residential intercept for Wasteville.

Step Six. Consider the sum of the products calculated in step four for the future period. Add this sum to the estimated intercept for the community, and the result is a forecast of the quantity of residential waste discarded per capita per day.

For Wasteville, the sum of the products for 1995 was 10.6875. This value plus the estimate of the intercept gives 3.27, which is the forecast for 1995 of the pounds of residential waste discarded in Wasteville per capita per day. This value is greater than the 1990 quantity by 0.7808, which suggests that without a user fee, residential waste will increase by approximately three quarters of a pound per person per day over the period from 1990 to 1995.

The forecasting method just outlined requires data for only one previous period. If data that represent both the quantity of waste and the independent variables are available for two or more previous periods, then there is a way to calculate an improved forecast. It involves re-estimating the intercept. The estimate calculated above is a poor one because it reflects not just a community's unique characteristics but also all of the random events that had an impact on the quantity of waste discarded in the previous period. An example of such a random event is a strong wind storm that blows down many trees and leads to a lot of yard waste. Under such circumstances, data for the previous period would reflect an unusually high quantity of waste. The data would also give a forecast that is unusually high. To water down the importance of such events, when data for two or more previous periods are available the improved forecast method should be used.

To do so, instead of applying the procedures outlined in steps one through four to just one previous period, apply them to all the previous periods for which data are available. Step five now becomes slightly more complicated. To estimate the intercept, subtract the sum of the products for each previous period from the per capita per day quantity of residential waste for that period. The average of these differences gives an improved estimate of a community's intercept. This improved estimate should now be used in step six to complete the improved forecast.

Steps one through six give an estimate of the number of pounds of residential waste discarded per capita per day in the forecast period. To calculate the total quantity of waste in the forecast period, simply multiply that estimate first by the size of the population and then by 365, unless, of course, it is a leap year. For example, for Wasteville the total quantity of residential waste expected in 1995 is approximately 6,900 tons, 1,900 more than in 1990.

Now, how big a difference would a user charge for residential SWS make? In general, to estimate the reduction in the per capita per day quantity of residential waste, simply multiply the user fee - the charge per 30- to 32-gallon container of refuse - by -0.40. To improve the estimate, deflate the user fee by an estimate of the consumer price index for the period in which the fee will be imposed.

Let us return to a consideration of Wasteville to illustrate how to predict the impact of a user fee. Recall that Wasteville officials needed two forecasts of the residential waste quantity: one assuming a continuation of their current policy of no user fee and one assuming a user charge of $1.00 per bag. Steps one through six gave an estimate of the former equal to 3.27 pounds per capita per day. To calculate the latter we should revise Table 8.2 so that the 1995 value for item 1 is $1.00 and then redo steps one through six. This leads to a final per capita per day quantity of residential waste in 1995 equal to 2.95 pounds. As expected, this is lower than the quantity estimated when a zero user charge was assumed, by the amount

0.32. This difference is simply the deflated user fee, $0.81, times -0.40.

As outlined above, without the $1.00 user fee the total quantity of residential waste in Wasteville is predicted to be 6,900 tons in 1995. This is almost 700 tons heavier than the quantity estimated when the user fee is assumed to be $1.00. Wasteville officials should consider this difference when deciding whether to charge a user fee. Of course, the decision should depend on other factors, too, such as the cost of implementing the program, the tipping fee that must be paid, and the welfare gain expected from a switch to residential user fees.

8.3 ESTIMATING THE WELFARE GAIN FROM A RESIDENTIAL USER FEE

A community should expect a welfare gain when it switches from a flat fee for residential SWS - or from a fee deducted from property taxes - to a user charge. This is true as long as the user charge reflects the social cost of SWS. By social cost we mean the sum of the private and public costs of SWS. An example of a private cost are the wages paid for the labor that collects refuse. An example of a public cost is the expected cost of possible ground water contamination by a landfill. Assuming that the user fee reflects such costs, the welfare gain from it can be estimated in two simple steps outlined below. To provide an example, we apply each instruction to the Wasteville data. Recall that Wasteville officials are considering a $1.00 user fee. We assume that this fee reflects the social cost of SWS in Wasteville.

Step One. Calculate the user fee per pound of SWS. To do this we need an assumption regarding the number of pounds of refuse that are typically in a 32-gallon container. Experiments in Seattle suggest that on average there are 20 to 22 pounds (Skumatz 1990, p. 12). Thus we assume a 32-gallon container holds 21 pounds of refuse. To calculate the user fee per pound of SWS simply divide the fee per 32-gallon container by 21.

For Wasteville, the fee per pound is $1.00 divided by 21, or approximately $0.05.

Step Two. Calculate the welfare gain per person per day. This requires multiplying the user fee per pound by the decline in waste discarded per person per day expected from the user fee. The resulting product should be divided by two because we are calculating the area of a welfare gain *triangle* (see Figure 1.1 in Chapter 1).

To calculate the gain per person per day for Wasteville we multiply $0.05 by 0.32, which gives 0.016. Dividing the product by 2 gives $0.008, which suggests that the user fee generates a welfare gain of almost a penny for each resident of Wasteville every day. This suggests an annual gain of almost $34,000 for the community.

8.4 FORECASTING THE QUANTITY OF COMMERCIAL WASTE

To forecast the quantity of commercial waste, data that represent the weight of commercial waste discarded in a previous period are needed. Data for the items in Table 8.3 for both the previous and the forecast periods are also needed.

Table 8.3 Data Required for a Community to Forecast Its Commercial Waste Quantity

---

1. User fee for commercial SWS (the weekly price in dollars per cubic yard for two pick-ups each week)

2. Mean temperature in degrees Fahrenheit

3. Total monthly precipitation in inches

4. Population density (population of the community divided by the square miles of land in the community)

5. Regional price paid for old corrugated containers per short ton

6. Employment in the community (the number of people employed by businesses located in the community; excluding federal, agricultural, mining, construction and manufacturing employment if the waste quantity data do not reflect the waste discarded by such businesses)

7. Producer price index divided by 100

Yearly averages or averages that relate to a specific month are acceptable for each data item with the exception of item 3. If monthly data is used then item 3 should be the total for the month rather than the average.

To see how the information in Table 8.3 is used let us return to Wasteville. Officials there estimate that the quantity of commercial waste was 4,000 tons in 1990. Assuming that 3,000 people worked there, there were 7.31 pounds of commercial waste per employee per day. Wasteville officials wish to forecast the quantity of commercial waste for 1995. They have succeeded in collecting data for 1990 and 1995 for each item listed in Table 8.3. These data are given in Table 8.4. The reference numbers that appear there correspond to those that appear in Table 8.3.

Table 8.4 Wasteville Data for the Commercial Forecast

	1990	1995
1.	$8.00	$9.00
2.	60	60
3.	2.5	2.5
4.	2,700	2,823
5.	$35.00	$34.00
6.	3,000	3,400
7.	1.18	1.20

Steps one through six below outline how to forecast the quantity of commercial waste.

Step One. Deflate the dollar value items in Table 8.4 by the producer price index; that is, divide items 1 and 5 by item 7. This calculation applied to Wasteville data leads to:

	1990	1995
1'.	$6.78	$7.50
5'.	$29.66	$28.33

Step Two. Divide item 4, population density, by 1,000. For Wasteville, we get:

	1990	1995
4'.	2.7	2.823

Step Three. For both the previous and the future time periods, multiply data items 1 through 5 by the corresponding coefficient estimated for the commercial equation. If the data have been revised by steps one and two, use the revised data. We reproduce the coefficients, which originally appeared in Table 6.1 below and associate each with a reference number as suggested by Table 8.3.

	Coefficient
1.	- 0.23
2.	0.02
3.	- 0.04
4.	- 4.31
5.	0.001

These coefficients multiplied by the corresponding 1990 and 1995 data for Wasteville give the following products.

	1990 Product	1995 Product
1.	-$1.5594	-$1.725
2.	1.2	1.2
3.	- 0.1	- 0.1
4.	-11.637	-12.1671
5.	$0.0297	$0.0283

Step Four. Sum the products obtained in step four for each time period. For Wasteville, the sums are -12.0667 for 1990 and -12.7638 for 1995.

Step Five. Consider the sum of the products for the previous time period only. Subtract this sum from the quantity of waste per employee per day that was discarded in the previous period. The difference is an estimate of the commercial intercept for a particular community. This intercept is analogous to the residential intercept discussed in Section 8.2. It reflects the characteristics of a community that affect the commercial waste discarded there and that are not represented by the five independent variables in the commercial equation.

Recall that the quantity of commercial waste discarded in 1990 in Wasteville was 7.31 pounds per employee per day. This value minus the sum of the products for 1990, -12.0667 yields an estimate of Wasteville's commercial intercept of 19.3767.

Step Six. Consider the sum of the products for the future period. This sum plus the estimated intercept gives a forecast of the quantity of commercial waste per employee per day for the future period. Often one will want to convert this to an annual forecast that represents the waste for the entire community and not just for a single employee. As the data for Wasteville will show, the amount of commercial waste discarded in a community may go up even though waste per employee goes down.

For Wasteville, the sum of the products for 1995 was -12.7638. This value plus the estimate of the commercial intercept, 19.3767, gives a forecast of the quantity of commercial waste discarded in Wasteville per employee per day of 6.61. This is over half a pound lighter than in 1990. Even so the total quantity of commercial waste discarded in Wasteville will not decline. In fact, we expect total commercial waste to change from 4,000 tons in 1990 to 4,100 tons in 1995. The reason is that the number of employees goes up by enough to counteract the decline in waste per employee.

There are two primary reasons for the decline in waste per employee: the expected increase in the commercial user fee and the expected increase in the population density. The latter is important because with a higher population density more service sector businesses are expected. Relative to other industries, the service sector probably generates low rates of waste per employee.

Steps one through six give a forecast of the quantity of commercial waste with a minimum of data. An improved forecast can be calculated if additional data are available. Data are required that represent the quantity of commercial waste and the independent variables listed in Table 8.3 not just for one previous period but for at least two. What makes this forecast better is explained in Section 8.2. Only now the random events that worry us are those that affect the quantity of commercial waste, such as a strike by employees.

Given access to the additional data, steps one through four above should be applied to each previous period rather than just to one. Then step five should be revised and the commercial intercept re-estimated. To re-estimate it, consider a single previous time period. Subtract the sum of the products for that period - as calculated in step four - from the quantity of commercial waste discarded per employee per day in the same period. Repeat this procedure for each previous period. Take the average of all the differences. The average is an improved estimate of the commercial intercept for the specific community.

The improved estimate represents a measure of the lump sum effect of the unique characteristics of a community that influence the quantity of commercial waste discarded there. It is better than the estimate calculated with data for just one previous period because it averages out any error term that might appear in the data for a single time period. Thus, if possible an improved estimate of a community's intercept should be estimated. Step six should then be followed using the improved estimate.

8.5 FORECASTING THE SUM OF RESIDENTIAL AND COMMERCIAL WASTE

Many communities do not keep separate track of residential and commercial waste quantities. These communities are limited to forecasting the quantity of combined waste for a future period. Such a forecast requires data for at least one previous period that represents the quantity of combined waste discarded in the community. It also requires data for the previous period and for the forecast period for the independent variables listed in Tables 8.1

and 8.3. Once these data are in hand the steps outlined below should be followed to forecast the combined quantity of commercial and residential waste for a given community.

To illustrate each instruction we shall apply it to data for the hypothetical community, Wasteville. These data are given in sections 8.2 and 8.4. Only now we assume that Wasteville officials do not know the separate quantities of residential and commercial wastes discarded in 1990. They only know that the sum of residential and commercial waste was 9,000 tons. This suggests that mixed waste per capita per day was 4.48 pounds. Wasteville officials also have data for 1990 and 1995 for the independent variables described by Tables 8.1 and 8.3 as specified in previous sections.

Step One. The same preliminary steps are required as if we were calculating separate forecasts of the quantity of residential and commercial waste. Thus, steps one and two in Sections 8.2 and 8.4 should be followed.

These instructions are applied to the Wasteville data and the results given in Sections 8.2 and 8.4 above.

Step Two. Consider the independent variables for the commercial equation only; that is, the variables described by Table 8.3. For each period, multiply data items 1 through 5, using revised data when it has been affected by step one above, by the ratio of employment to population. Employment is data item 6 in Table 8.3 and population is data item 9 in Table 8.1.

Wasteville's commercial data items 1 through 5 are given below.

Commercial Data Items

	1990	1995
1'.	$ 6.78	$ 7.50
2.	60	60
3.	2.5	2.5
4'.	2.7	2.823
5'.	$29.66	$28.33

The employment to population ratio for Wasteville is 3,000/11,000 or 0.2727 for 1990 and 3,400/11,500 or 0.2957 for 1995. Multiplying each of the above data items by the applicable ratio gives the following:

<u>Commercial Data Items after Multiplication</u>
<u>by the Ratio of Employment to Population</u>

	<u>1990</u>	<u>1995</u>
1'.	$ 1.8489	$ 2.2178
2.	16.362	17.742
3.	0.6818	0.7393
4'.	0.7363	0.8348
5'.	$ 8.0883	$ 8.3772

Step Three. Apply step three which is explained in Section 8.2, to the residential data. Apply step three in Section 8.4 to the commercial data, being careful to apply it to the commercial data items as revised in step two immediately above. That is, multiply commercial data items 1 through 5, after revising them as specified immediately above, by the corresponding coefficients estimated for the commercial equation, and multiply residential data items 1 through 8 by the corresponding coefficients estimated for the residential equation.

We obtain the following products for Wasteville by multiplying the residential data items by their corresponding coefficients.

<u>Residential Sector</u>

	<u>1990 Product</u>	<u>1995 Product</u>
1.	$ 0	$ 0
2.	$ 0.7025	$ 0.7662
3.	0.6	0.6
4.	0.1	0.1
5.	- 7.3975	- 7.095

6.	6.072	6.039
7.	9.828	10.2757
8.	$ 0.0017	$ 0.0016

These products are identical to those obtained in Section 8.2. The products for the commercial sector are different. We obtain the following products by multiplying the revised commercial data items by their corresponding coefficients.

Commercial Sector

	1990 Product	1995 Product
1.	-$ 0.4252	-$ 0.5101
2.	0.3272	0.3548
3.	- 0.0273	- 0.0296
4.	- 3.1735	- 3.598
5.	$ 0.0081	$ 0.0084

Step Four. Sum the residential and commercial products obtained in step three for each time period.

For Wasteville, the sum for 1990 is 6.616; for 1995 it is 6.913.

Step Five. Consider the sum for the previous period only. Subtract this sum from the quantity of combined waste discarded per capita per day in the previous period. The difference is an estimate of the sum of the residential and commercial intercepts for a community. The intercept reflects the characteristics of a community that might affect the quantity of residential or commercial waste discarded there but that are not represented by the independent variables. If possible this intercept should be estimated using data for more than one previous time period. Otherwise any random occurrences in a single previous time period that affected the quantity of waste discarded will skew the intercept estimate. Thus, if possible, follow steps one through four for several previous periods, and then calculate the difference between the sum of the products and the actual quantity discarded in the previous periods. An improved estimate of the intercept would be the average of the differences.

For Wasteville the difference is obtained by subtracting the sum of the products for 1990, 6.616, from the quantity of mixed waste discarded per capita per day in 1990, 4.48. This gives -2.136. Thus an estimate of the sum of the residential and commercial intercepts for Wasteville is -2.136.

Step Six. Consider now the sum of the products in step four for the future period. Add this to the estimated intercept for the community and the result is a forecast of the sum of residential and commercial waste discarded per person per day in the forecast period.

Recall that for Wasteville the sum of the products in step four for the future period was 6.913. Thus the estimate of the quantity of mixed waste for 1995 is 4.777 or the sum of the intercept and 6.913. This is greater than the quantity discarded in 1990 by 0.3 pounds per person per day.

8.6 SOME FINAL REMARKS REGARDING FORECASTS

Perhaps the most difficult aspect of forecasting is locating good estimates of the future values of the independent variables. Regional economic development or planning offices are often good sources. For other ideas for sources see Table 5.4, which lists the sources used by the author for the sample communities. Any errors in the future values of the independent variables lead to errors in the forecasts of waste quantities. Thus, special care should be taken when collecting these data.

9. Summary and Conclusions

9.1 AN OVERVIEW

Many US and European communities are now facing a shortage of waste disposal capacity. One way to dampen the demand for such capacity is to charge a user fee for residential SWS. User fees for SWS are commonly charged to the commercial sector, but they have rarely been charged to households, at least until recently, when solid waste officials in many communities began experimenting with such user fees. The stated reason for these 'experiments' is to decrease the demand for SWS. An important question, then, is how big is the impact of user charges on the demand for both commercial and residential SWS.

The existing literature concerning such an impact is quite sparse. The few studies undertaken have lacked empirical rigor, often because of limited data.

Residential demand for SWS can easily be modeled within a utility maximization framework; commercial demand can be modeled within a profit maximization framework. We develop two such models. The utility maximization model suggests that the time the household devotes to recycling will increase with a user fee for SWS. It also suggests that as long as the household is paying a user fee, the cost of a good to the household will increase with the quantity of waste associated with that good. Not surprisingly, the profit maximization model suggests analogous conclusions for the firm. Firms will devote more labor to recycling and will find inputs that generate a lot of waste more expensive as the commercial user fee increases.

The models we develop also suggest that certain variables affect the demand for residential and commercial SWS. We collect data that represent these variables - one of which is the user charge for SWS - and that represent the quantity of waste discarded for nine communities in the US. Five of these communities charge residential user fees for SWS; four do not. The data collected are monthly and cover a period of at least one year for each community. We construct a panel data set by pooling the data for the nine communities. The nature of the data that represent the quantity (tons) of waste discarded suggests that the error structure of our empirical model

127

is heteroskedastic.   Thus, our estimation technique is generalized least squares (GLS).

The tonnage and price data for each of the nine communities were collected by directly contacting the municipalities. As a result, the data are marked by inconsistencies. For example, the categories of waste measured by the tonnage data vary from community to community. Also, the laws and institutions that might affect the demand for SWS vary from community to community.

The results of our GLS estimation suggest that the response of the household and the firm to user fees for SWS is significant. In particular, if a community switches from a flat fee for SWS or from a fee that is deducted from property taxes to a user charge of $1.00 per 32-gallon container, its waste should decline by nearly a half pound per person per day. The average quantity of residential waste discarded in the sample communities is 2.60 pounds per capita per day. Thus a switch to a $1.00 user fee would lead to a reduction of approximately 15 percent in residential waste.

To determine the robustness of our results, we reformulate the empirical model in several ways. First, we create an annual model that corresponds to the original monthly one. Estimation of the annual model produces coefficients similar to those of the monthly model. Second, we employ a Hausman test of endogeneity to verify that the user fee variables are exogenous. The results suggest that our assumption of exogeneity is correct. Finally, we test whether the slope coefficients of the model are constant across communities. The results of this test suggest that the slope coefficients do vary significantly over communities. However, because of the small number of observations and the extent of multicollinearity that relates to each individual community, the coefficients estimated separately for each community are almost always insignificant. Thus, while pooling the data does introduce some bias, this bias may be offset by the large reduction in mean squared errors that should result from an increased sample size.

The demand models estimated for residential and commercial SWS may be used to forecast the quantity of waste discarded by a particular community. Forecasting requires two types of information: data that represent the quantity of waste discarded in a previous period and data that represent the independent variables, such as the average household income, for the previous period and for the future (forecast) period. We provide detailed instructions for forecasting the quantity of residential waste, the quantity of commercial waste and the welfare gain that a community should enjoy from switching to a user charge for residential SWS.

9.2 THE IMPLICATIONS FOR POLICY

The results of our empirical analysis suggest that the demand for SWS is sensitive to user fees. This holds for both residential and commercial demand. We estimate that the price elasticity of residential demand for SWS is -0.12 and that the price elasticity of commercial demand is -0.29. The elasticities suggest that the impact of an increase in the user fee for SWS on the demand for SWS by an entire community for a period of, say, one year is large enough that there could be a substantial reduction in municipal spending for SWS. Consider, for example, a switch to a $1.00 residential user fee for SWS in a typical US community with a population of 500,000. We would expect the quantity of waste discarded per person per day to decline by 0.4 pounds. This suggests a reduction of 36,500 tons in the quantity of waste discarded by the entire community during a year. If the community pays a reasonably low tipping fee such as $40.00 per ton, savings in disposal costs alone would be nearly $1.5 million. Once we take transportation and collection costs into consideration, savings would be substantially larger.

We calculate a range of estimates of the welfare gain to society that should result from imposing a residential user fee. The estimates are made assuming that the user fee is set equal to the marginal social cost of SWS. The welfare gain estimates vary directly with the size of the affected population and the estimate of the social cost of SWS. The lowest annual welfare gain estimated, $36,000, was to a small community with a very low social cost of SWS. The highest annual welfare gain estimated, $7,700,000, was to a community with the opposite characteristics.

Our conclusions lead to two suggestions regarding the policy options for extending the life of a local landfill. One is to stop financing residential SWS by flat fees or through property taxes and to instead charge user fees. The other is to raise the user fees for commercial SWS.

The value of these policy suggestions depends primarily upon two issues. The first concerns the administrative costs of imposing a residential user fee. The five communities in our sample that charged user fees never mentioned that these costs were prohibitive. In fact, some western communities in the US have charged user fees for over 50 years. These observations suggest that the administrative costs of a residential user fee for SWS can be low enough to compete with the cost of administering a flat fee program or of deducting the charge from property taxes. The two user fee variants represented by the sample communities are certainly administratively feasible. One variant is to sell tags or plastic bags that are marked with a distinctive logo and that are priced to recoup the eventual collection and disposal costs associated with 32 gallons of refuse. This variant does not require customer billing. The bags or tags are made available through local retail establishments, which are usually eager to offer them because they draw customers into the store. The other user fee variant is to have residents subscribe to a certain number of cans of service per week. This requires billing customers but often the bill for SWS is simply appended to a water or other utility bill. Clearly, the administration of either variant is practical and may not be prohibitively expensive. The cost, however low, should be weighed against the value of the reduction in waste that user fees will bring. The cost might also be weighed against the value of the fairness that user fees bring by making the charge for SWS vary directly with the quantity of waste discarded.

A second issue that affects the value of our policy suggestions is the magnitude of the increase in illegal disposing that accompanies a user fee program. Two of the communities in our sample, Seattle and San Francisco, suggested that this was an ongoing concern. Other communities[1] have experienced surges in illegal dumping during the initial months of their user fee program, but they have found that quick and consistent enforcement of anti-dumping laws soon dampens the illegal response. A common tactic of rebellious households is to deposit their refuse in nearby commercial dumpsters. Locking such dumpsters alleviates the problem but increases the time that haulers spend collecting refuse since the dumpsters have to be unlocked before they are emptied. Another tactic of households is to transfer waste to a relative's home where unlimited SWS are available. One enforcement method that several communities found useful was for the refuse collectors to keep an eye out for households that consistently discard very little or no refuse. Officials suggest that such households be issued strong warnings against illegal disposal practices.

Exactly how big the illegal response to user fees is will certainly vary from community to community. Low-income communities seem prime candidates for a big illegal response since the user fee for SWS will be a relatively large budget item there and thus worth some effort to avoid. Rural areas also seem susceptible since they contain many open spaces where dumping might not be detected for some time. In general those communities where dumping seems unlikely are those that are environmentally active or that have a track record of cohesiveness on other fronts. Certainly, however, the relationship between illegally discarded quantities and user fees for SWS is a topic in need of further research.

NOTES

1. A community north of Chicago and several communities in upstate New York experienced such surges.

Appendix A

Table A.1 The Residential and Commercial User Fees by Community Over Time				
Community	Residential User Fee[a]		Commercial User Fee[b]	
San Francisco	$0.69	1/80 - 6/80	$ 9.66	1/80 - 6/80
	0.86	7/80 - 6/82	11.99	7/80 - 6/82
	1.01	7/82 -10/83	14.10	7/82 -10/83
	1.17	11/83 - 6/87	16.26	11/83 - 6/87
	1.26	7/87 - 9/88	18.86	7/87 - 9/88
Hillsborough County	0	1/84 - 6/84	6.90	1/84 -12/84
		8/84 - 2/88	7.12	1/85 -12/85
			7.58	1/86 -12/86
			8.49	1/87 -12/87
			9.09	1/88 - 2/88
St. Petersburg[c]	0	1/80 -12/88	--	
Estherville	1.00	7/84 - 6/85	10.23	7/84 -11/88
	0.25	7/85 -11/88		
Howard County[c]	0	7/87 - 7/89	--	
Highbridge	1.73	1/88 -12/88	10.31	1/88 -12/88
Bernalillo County	0	7/82 - 5/89	2.36	9/82 - 2/84
			2.71	3/84 - 9/84
			2.85	10/84 - 3/87
			3.15	4/87 - 3/88
			3.97	4/88 - 5/89

Seattle[c]	0	1/80 -12/80		
	0.50	1/81 - 6/82		
	0.67	7/82 -12/84	--	
	0.73	1/85 - 7/86		
	1.31	8/86 - 5/87		
	1.71	6/87 -12/88		
Spokane	0.70	1/86 -12/86	7.85	1/86 -12/86
	0.73	1/87 -12/87	8.18	1/87 -12/87
	0.98	1/88 -12/88	11.08	1/88 -12/88

a. The residential user fee is the weekly charge per 30- to 32-gallon container.
b. The commercial user fee is the weekly charge per cubic yard for two pick-ups a week.
c. The data for this community relate only to residential waste quantities.

Table A.2 Mean Values of Independent Variables by Community (Part 1)[a]				
Variable	San Francisco	Hillsbor- ough County	St. Petersburg	Estherville
Age Distribution of the Population	52.9%	49.0%	37.1%	42.5%
Household Income[b]	$23,963	$21,580	$17,087	$16,399
Household Size	2.24	2.69	2.25	2.48
Price of Old News- papers[b]	$35.47	$27.98	$32.85	$34.45
Price of Old Corrugated Containers[c]	$50.80	$41.18	--	$36.98
Population	715,013	462,999	241,888	6,479
Employ- ment	467,041	74,475	--	1,670
Square Miles	44.6	933.0	55.7	4.6
Precipi- tation	1.86	3.50	3.89	2.36
Average Temper- ature	58.55°	72.66°	72.50°	46.31°
Diesel Fuel Prices[c]	$43.68	$60.56	$45.23	$56.11

Wages of Workers in Refuse Systems[c]	$19.89	$24.21	$18.83	$16.46
	n=99	n=43	n=102	n=47

Table A.2 Mean Values of Independent Variables
         by Community (Part 2)[a]

Variable	Howard County	Highbridge	Bernalillo County	Seattle
Age Distribution of the Population	56.9%	50.8%	52.2%	52.4%
Household Income[b]	$36,552	$34,834	$20,063	$22,636
Household Size	2.77	2.90	2.63	2.17
Price of Old Newspapers[b]	$18.06	$13.03	$38.05	$36.44
Price of Old Corrugated Containers[c]	--	$23.78	$60.90	--
Population	165,837	4,200	471,469	492,166
Employment	--	334	168,890	--
Square Miles	251.0	2.5	1,169.0	83.6
Precipitation	4.87	4.81	0.88	3.12
Average Temperature	55.46°	59.25°	55.76°	53.49°
Diesel Fuel Prices[c]	$46.16	$42.09	$59.64	$43.56

Wages of Workers in Refuse Systems[c]	$25.50	$30.09	$18.77	$19.42
	$n=19$	$n=6$	$n=77$ or 75	$n=102$

Table A.2 Mean Values of Independent Variables by Community (Part 3)[a]	
Variable	Spokane
Age Distribution of the Population	46.8
Household Income[b]	$16,095
Household Size	2.29
Price of Old Newspapers[b]	$44.88
Price of Old Corrugated Containers[c]	$77.85
Population	171,460
Employ-ment	72,972
Square Miles	55.70
Precipita-tion	1.39
Average Tempera-ture	48.81°
Diesel Fuel Prices[c]	$46.66

Wages of Workers in Refuse Systems[c]	$24.28
	$n=30$

a. Means have been calculated after dropping the first six observations for each community. Recall that the final data set included two lagged variables whose construction required omitting the first six observations for each community.
b. Data have been deflated by the CPI.
c. Data have been deflated by the PPI.

Appendix B

Table B.1 List of Sources of Data Related to Community Characteristics and Regional Prices

1. Consumer Price Index: US Department of Labor, Bureau of Labor Statistics, Washington, DC.

2. Producer Price Index: US Department of Labor, Bureau of Labor Statistics, Labstat Series Report, Washington, DC.

3. Age Distribution of the Population, Household Income, Household Size: *1981 through 1989 Survey of Buying Power*, Published by Sales and Marketing Management, New York, NY.

4. Price of Used Newspapers, Price of Used Corrugated Containers: Official Board Markets, *The Yellow Sheet*, Edgell Communications, Chicago, Illinois.

5. Population:

    San Francisco - California Department of Labor, Employment Development Research.

    Hillsborough County - Hillsborough Planning Commission.

    St. Petersburg - St. Petersburg's Planning Commission.

    Estherville - Iowa State Department of Employment Service, Des Moines, Iowa.

    Howard County - Howard County Department of Planning and Zoning, Residential Construction and Population Report.

Highbridge - New Jersey State Department of Labor, Demographic and Economic Analysis.

Bernalillo County - New Mexico Department of Labor.

Seattle - Seattle City Light Utility.

Spokane - Spokane Planning and Zoning Commission.

6.   Employment: US Department of Labor, Bureau of Labor Statistics, Reports and Analysis, Employment, Wages and Contributions, ES202.

7.   Square Miles: US Bureau of the Census

8.   Precipitation and Average Temperature: Local Climatological Data, National Climatic Data Center, US Department of Commerce, Asheville, North Carolina.

9.   Diesel Fuel Prices: No. 2 Distillate Wholesale Sales, Department of Energy, Energy Information Administration, Washington, DC.

10. Wages of Workers in Refuse Systems: Table 5, 4953 Refuse Systems, Employment and Wages, Annual Averages 1982 - 1989, US Department of Labor, Bureau of Labor Statistics, Washington, DC.

References

Bagby, Jennifer. (1988) 'Waste Reduction, Recycling and Disposal Alternatives, Appendix E - Recycling Potential Assessment and Waste Stream Forecast' Report for the Seattle Solid Waste Utility, Seattle: Seattle Solid Waste Utility, July.

Baumol, William J. (1977) 'On Recycling as a Moot Environmental Issue' *Journal of Environmental Economics and Management*, 4, March, pp. 83-7.

Beck, R.W. and Associates. (1987) *The Nation's Public Works: Report on Solid Waste.* National Council on Public Works Improvement, May.

Bish, Robert L. and Nourse, Hugh O. (1975) *Urban Economics and Policy Analysis*, McGraw-Hill Book Co., New York.

Cargo, Douglas B. (1978) *Solid Wastes: Factors Influencing Generation Rates*, The University of Chicago, Chicago.

Church, George J. (1988) 'Garbage, Garbage Everywhere' *Time*, 5 September.

Efaw, Fritz and Lanen, William N. (1979) 'Impact of User Charges on Management of Household Solid Waste' Cincinnati Municipal Environmental Research Lab, prepared by Mathtech, Inc., Princeton, NJ, August.

Franklin Associates, Ltd. (1988) *Characterization of Municipal Solid Waste in the U.S., 1960 to 2000*, Franklin Associates, Ltd., Prairie Village, Kansas, March.

Glebs, R.T. (1988) 'Subtitle D: How Will it Affect Landfills?' *Waste Alternatives*, March.

Hanley, Robert. (1988) 'Pay-by-Bag Trash Disposal Really Pays, Town Learns' *New York Times*, 24 November.

Jenkins, Robin R. (1991) 'Municipal Demand for Solid Waste Disposal Services: The Impact of User Fees' Unpublished PhD Dissertation, University of Maryland, College Park, Maryland.

Johnston, J. (1984) *Econometric Methods*, McGraw-Hill Book Co., New York.

Kemper, Peter, and Quigley, John M. (1976) *The Economics of Refuse Collection*, Ballinger Publishing Co., Cambridge, Mass.

Langham, Max R. (1972) 'Theory of the Firm and the Management of Residuals' *American Journal of Agricultural Economics*, 54, pp. 315-22.

McFarland, J.M. et al. (1972) 'Comprehensive Studies of Solid Waste Management' Report for the US Environmental Protection Agency, The University of California, Berkeley, May.

Menell, Peter S. (1990) 'Beyond the Throwaway Society: An Incentive Approach to Regulating Municipal Solid Waste' *Ecology Law Quarterly*, University of California, Berkeley, California, 17, 4.

Miedema, Allen K. (1983) 'Fundamental Economic Comparisons of Solid Waste Policy Options' *Resources and Energy* 5, pp. 21-43.

Neal, Homer A. and Schubel, J.R. (1987) *Solid Waste Management and the Environment: The Mounting Garbage and Trash Crisis*, Prentice-Hall, Inc., Englewood Cliffs, New Jersey.

Official Gazette of the City of Spokane, Washington (1989) 4 January.

Organization for Economic Co-operation and Development. (1981) *Economic Instruments in Solid Waste Management*, Organization for Economic Co-operation and Development, Paris.

Platt, Brenda A., et. al. (1988) *Garbage in Europe: Technologies, Economics, and Trends*, Institute for Local Self-Reliance, Washington, DC, May.

Prodromidis, Kyprianos P. and Lianos, Theodore P. (1973) 'City Size and Patterns of Employment Structure' In *Proceedings of the Conference on Urban Economics*, edited by John M. Mattila and Wilbur R. Thompson, Wayne State University, Detroit.

Rathje, William L., Reilly, Michael D. and Hughes, Wilson W. (1985) 'Household Garbage and the Role of Packaging' Report for the American Paper Institute, Prepared by Le Projet du Garbage, University of Arizona, Tucson.

Rathje, William L. and Thompson, Barry. (1981) 'The Milwaukee Garbage Project' Report for the Solid Waste Council of the Paper Industry, Prepared by Le Projet du Garbage, University of Arizona, Tucson, March.

Richardson, Robert A. and Havlicek, Joseph Jr. (1975) 'An Analysis of the Generation of Household Solid Wastes From Consumption' Research Bulletin, No. 920, Purdue University, West Lafayette, Indiana, March.

Richardson, Robert A., and Havlicek, Joseph Jr. (1978) 'Economic Analysis of the Composition of Household Solid Wastes' *Journal of Environmental Economics and Management*, 5, pp. 103-11.

Savas, E.S. (1977) *The Organization and Efficiency of Solid Waste Collection*, D.C. Heath and Company, Lexington, Mass.

Schmidt, Peter. (1976) *Econometrics*, Marcel Dekker, Inc., New York.

Skumatz, Lisa A. (1990) 'Volume-Based Rates in Solid Waste: Seattle's Experience' Report for the Seattle Solid Waste Utility, Seattle Solid Waste Utility.

Stevens, Barbara. (1977) 'Pricing Schemes for Refuse Collection Services: The Impact on Refuse Generation' Unpublished, Research Paper #154, Graduate School of Business, Columbia University, New York.

Stevens, Brandt K. (1988) 'Fiscal Implications of Effluent Charges and Input Taxes' *Journal of Environmental Economics and Management*, 15, September, pp. 285-96.

Theil, Henri. (1971) *Principles of Econometrics*, John Wiley & Sons, New York.

US Congress. (1989) Office of Technology Assessment, *Facing America's Trash: What Next for Municipal Solid Waste*, OTA-0-424, US Government Printing Office, Washington, DC, October.

US Department of Labor. (1988) Bureau of Labor Statistics, *Handbook of Methods*, Bulletin 2285, Washington, DC.

US Environmental Protection Agency. (1987) Office of Solid Waste, *Draft Final Report: Survey of State and Territorial Subtitle D Municipal Landfill Facilities*, submitted by Westat, Inc., October.

US Environmental Protection Agency. (1988) Office of Solid Waste, *Draft Regulatory Impact Analysis of Proposed Revisions to Subtitle D Criteria for Municipal Solid Waste Landfills*, prepared by Temple, Barker & Sloane, Inc., August.

Wertz, Kenneth L. (1976) 'Economic Factors Influencing Households' Production of Refuse' *Journal of Environmental Economics and Management*, 2, April, pp. 263-72.

1988 Revisions to San Francisco County Solid Waste Management Plan. (1988), March.

Index

Air emissions 43
Apartment waste 66
Bernalillo County, New Mexico 63, 66, 67, 71
Bottle bill 66, 68
Cargo, Douglas B. 28
Commercial Waste 6, 66, 67, 69, 86, 110, 117, 121
      Decline in waste per employee 120
      Service sector 120
Compacting refuse 19
Compost 33, 85
Connecticut 23
Construction and demolition debris 69
Contamination of water by landfills 3, 90, 92, 116
Conversion of gallons of refuse to pounds 116
Cost minimization model 50
Cost of SWS
      Avoidance of 38, 45, 49
      Constant returns to scale 91
      Landfilling 92
      Marginal cost of SWS 1, 3, 4, 90
      Private cost of SWS 1, 4, 90, 116
      Public cost of SWS 1, 4, 90, 116
      Social cost of SWS 1, 4, 90, 93, 116, 129
Cost of waste disposal
      And user fees 129
      Land 3
      Landfilling 92
      Public aversion to disposal sites 3
      Stricter EPA regulations 3, 93
      Tipping fee 89, 129
Data that represent community characteristics 62, 72, 124